I was fortunate in having considerable dealings in 1938–40 with the 'Boffins' (as the Royal Air Force affectionately dubbed the scientists) and will never forget the invigorating atmosphere that pervaded Bawdsey and later, Worth Matravers. Here was brilliant individualism harnessed to make a great team without loss of individual freedom and initiative … It is perhaps because this type of teamwork is of the very essence of the Royal Air Force that such happy relations were established between the Service and our friends and partners, the Boffins … Such a team spirit might well never have been created had it not been for men like Rowe and Tizard who could and did interpret the user to the scientist and the scientist to the user. They, and men like them, set a pattern which is not only vital for our future security, but may well, I suggest, be vital for the well-being of the world.

Marshal of the RAF Lord Tedder, 1948

SECRET LOCATION

A Witness to the Birth of Radar and its Postwar Influence

IAN GOULT

The History Press

First published 2010

The History Press
The Mill, Brimscombe Port
Stroud, Gloucestershire, GL5 2QG
www.thehistorypress.co.uk

British Library Cataloguing in Publication Data.
A catalogue record for this book is available from the British Library.

ISBN 978 0 7524 5776 5

Typesetting and origination by The History Press
Printed in Great Britain
Manufacturing managed by Jellyfish Print Solutions Ltd

Contents

Acknowledgements

Thanks are due to the following for their help and support. Shaun Barrington of The History Press for his continuous encouragement. Dr Bill Penley and Dr Phil Judkins, Vice Chairman and Chairman of the Purbeck Museum Trust for permission to use photographs and images from *Dorset Radar Days*, and Dr Judkins for advice on copyright. Mary Wain, Chairman of the Bawdsey Museum Trust for permission to include photographs relating to Bawdsey Manor. The late Mrs Keith Wood and her daughter Judith for permission to include Keith Wood's description of the historic first air-to-air radar contact from his book *Echoes and Reflections*. Also, I would like to thank author Siobhan Curham, Leader of the Uxbridge Library Writers Group for her encouragement and to the friendly staff at Ruislip Library during my frequent visits there; and my sister Joan for securing some of the above contacts. Finally I am grateful for the patience and forebearance of my wife during the preparation of the manuscript.

PROLOGUE

Early Days – Introducing the Cat's Whisker

I was born in 1924 in the front room of my grandparents' hotel, within the sound of waves breaking on the beach at Felixstowe. My cousin's parents also ran a nearby hotel. Apart from messing about in boats we played with early wireless sets. It was possible to listen to the 'long wave' programme with a piece of silicon crystal, a cat's whisker (a short steel spring pushed against the silicon chip) a coil of wire, and a pair of headphones. I built my first working radio at the age of twelve, and experimented with 'short wave' radio – HF in today's terminology. This early experience was to prove useful later.

In retrospect a great opportunity was missed in these early days of radio. Crystal sets were considered the province of amateurs pursuing a hobby, which was true. Had the physicists of the day deigned to investigate the reason why the silicon junction acted as a radio signal detector, the transistor and associated integrated circuits (silicon chips) would have evolved 50 years earlier. The era of thermionic valves used for early radios and televisions would have been by-passed. I was to come across the 'cat's whisker' silicon diode, properly engineered, some six or seven years later when working on microwave radars, where thermionic diodes were unsuitable. Some of the most prominent physicists of the day were involved in these early microwave radars, but had more pressing problems on their minds than indulging in fundamental research on why the silicon diode worked as it did – again delaying the semi-conductor and digital computer era. So my early wireless sets taught me about valve technology. I did not envisage learning a new technology twenty years later.

At the outbreak of war in September 1939 the hotels were requisitioned to house the massive intake of recruits undergoing military training. My parents were left with virtually no income – or home. My dreams of staying on at school and continuing to university were finished.

Bawdsey Manor was situated on the opposite side of the river Deben from Felixstowe. Unknown to the general public this was where government scientists were evolving what was later to be known as RADAR. My sister worked there as a secretary. As I was too young for national service my sister suggested that I should apply for a job at the establishment. I knew from the aerial towers that it had something to do with radio so I readily agreed.

I joined the Telecommunications Research Establishment (TRE), as it was later called, at Swanage, where it was moved to in the early days of the war on the basis that it was a more secure location. I joined in 1940 shortly before the Fall of France and the evacuation of our forces from Dunkirk.

PART I

The Telecommunication
Research Establishment
– TRE

CHAPTER I

How it all Started
– Orfordness and
Bawdsey Manor

The story that follows is not the subject of detailed research. It is based on the author's memory and first-hand experience of working at TRE during the Second World War. There may be inaccuracies of detail due to the passage of time, but the general background is as authentic as memory permits. Events that I was unaware of but came to light later are dealt with in the third person.

During the early 1930s whilst I was still playing with crystal sets, Watson-Watt, a government scientist, was working on the investigation of radio 'noise' and its sources. The popular press of that time were carrying stories of a 'death ray' that could disable aircraft and anything else they could think of. There was no evidence for this, but it made a good story and began to be taken seriously. Questions were asked at the highest level.

In 1935 Watson-Watt was asked to investigate the feasibility of such a death ray. The investigation turned on its head. One of his scientists, A. V. Wilkins, calculated that it was not feasible, but suggested that aircraft would interfere with the transmission of radio waves. The possibility led to an experiment using an aircraft to fly through the transmission from the BBC transmitter at Daventry. This confirmed Wilkin's suggestion, and made possible, in principle, the detection of aircraft by the reflection of radio waves. The importance of this experiment was immediately recognised by the parliamentary committee that had been set up to look into the country's air defence. Funding was made available for further research.

Orfordness

Orfordness must be one of the oddest locations in the UK, subject to wild rumours down through the decades and if anything, the site of even wilder actual events. Hidden away on the East Suffolk coast, the Ness is now a National Trust nature reserve – but it is also a military experimental site. The military occupied what the locals call 'the Island', a stretch of marsh and shingle on the far side of the river Ore, for 80 years.

To give just one example of the kind of things that went on there, in April 1942, the A.R.P. asked for help in supplying guidelines to the building industry for constructing roofs capable of withstanding German 1 kg incendiaries. The story is told in the excellent book *Most Secret: The Hidden History of Orford Ness* by Paddy Heazell.

In order to research the typical angle of hit, depth of penetration and concentration of spread from a standard stick of bombs, they decided to go for complete replication of a bombing raid even to the extent of using a captured

German aircraft. It was armed with loads of retrieved unexploded bombs. RAF Duxford kept such aircraft in operational order and supplied a Junkers Ju 88A. This particular one had fallen into RAF hands the previous July by a neat trick. Its pilot was totally bamboozled by false radio signals and much to his surprise, he found himself landing at Lulsgate, near Bristol, when imagining he had reached his Brittany base. The bombs, all repaired, had their explosive charges replaced by plaster of Paris, suitably weighted with steel shot, to replicate in all respects the weight and balance of the real thing. They were then dropped over King's Marsh from the two usual operational heights, 3,000 ft and 7,000 ft. The aim was to analyse every single bomb strike. In the event about 70% were traceable and these supplied sufficient evidence for the purpose. A 2-inch mortar was also used to simulate the impact onto specific surfaces. The results were tabulated, bomb by bomb. It makes turgid reading, but at a grim period in the war, the issue was highly charged. The A.R.P. was pleased to be given the facts. The Ness's skills and usefulness were well demonstrated, not for the first or last time.

It was at Orfordness and Bawdsey Manor on the Suffolk coast that Watson-Watt set up the research establishment leading to the development of RDF (Radio Direction Finding), the early name for radar.

First, an encapsulated history of Bawdsey Manor: the Manor was built in 1886 and enlarged in 1895 as the primary residence for Sir Cuthbert Quilter. During the First World War the grounds and stables were requisitioned by the Devonshire Regiment. After the war the Bawdsey Manor

Estate was selected in 1935 as the site for a new research station for the development of radio direction finding. The Treasury allocating one million pounds for the continuation of the research started at Orfordness. The Manor, estate buildings and 168 acres of land were sold to the Air Ministry in 1936 and Watson-Watt (a direct descendant of James Watt inventor of the steam engine) was appointed as superintendent and began his work.

In January 1937 the RAF's Radio Direction Finding (RDF) training school was established there and the very first of the Chain Home radar stations was created there, coming on line in May 1937.

In August a filter room was established to process data from two other Chain Home stations. The station was operational by 24 September 1937 providing long range early warning for the Channel approaches and the southern North Sea, as well as radar for coastal convoys. Another important area of research was the development of an 'Identification, Friend or Foe' (IFF) system. Aircraft were fitted with aerials with motor-driven tuners that caused the reflected signal received by ground stations to vary in amplitude. Later models used an electronic unit that detected the presence of friendly radar and then transmitted a coded signal to indicate a friendly aircraft. By Easter 1939 15 Chain Home stations were established and running around the coast and Chain Home was on 24-hour watch.

Bawdsey continued in the forefront of the expansion of the radar network with an AMES Type 2 Chain home Low on a 200-foot platform on the southern (No 4 of 4) transmitter mast. Each mast was a staggering 360 feet high, giving it the ability to detect low flying aircraft and coastal shipping. Towards the end of 1941, Coastal Defence Radar

(Army CD Mk IV) was established at Bawdsey Manor. This installation was taken over by the RAF on 7 December 1942 making Bawdsey the only site in the UK with three types of radar (CH, CHL and CD) in operation. By August 1943 Coastal Defence was changed to an AMES Type 55 Chain Home Extra Low (CHEL). In September 1944 the Manor began monitoring the V2 rockets using specially developed Chain Home receivers codenamed 'Oswald'. Oswald indicated to Bomber Command the location of the launch sites, which could then be attacked. Bawdsey was still operating both CH and CHEL in 1948.

So, to return to the outbreak of war, the whole of the south coast and much of the east coast was covered by a chain of RDF stations that could detect the approach of aircraft up to a distance of 200 miles, thus giving early warning of raids, enabling the scrambling of fighters to intercept and engage the enemy.

The technique was to transmit a high power radio signal of short duration (typically 4 microseconds) and observe the time delay of the echo reflected back from the aircraft on a calibrated time base of a cathode ray tube. Radio waves travel at the speed of light, 186,000 miles per second; thus the go and return time as indicated on the cathode ray tube represents the distance of the aircraft from the transmitter. Receivers using some of the techniques employed for television receivers formed the basis of the echo detectors.

The technology was new and developed by scientists rather than engineers. As there was no precedence for the installation, setting up and commissioning of the early RDF stations, the academic research staff from Bawdsey found themselves operating on building sites.

Who was to operate and interpret the blips on the RDF

cathode ray tubes? Three typists from Bawdsey, one the wife of a senior scientific officer, were given rudimentary instruction and set up as guinea pig operators. They were able to correctly interpret the information presented on the cathode ray tubes. This led to the training of WAAFs and WRNS as radar operators.

Following the success of the ground Radio Direction Finding system, attention was given to the development of an airborne system for Air to Surface Vessel detection (ASV). How could the massive transmitters and receivers, each situated in their own building with 360 foot aerial masts, be fitted to aircraft? The answer lay in the increase of the operating frequency, which would result in a reduction of the wavelength of the transmitted signal enabling the use of smaller, lighter equipment and aerials.

The ground-based chain of RDF stations transmitted at frequencies between 30–45MHz; wavelengths of 10–15m. Trials conducted as early as 1937 indicated that a specially designed lightweight pulse transmitter/receiver operating at a wavelength of 1.25m (240MHz) with an aerial system similar to that of early television receivers could locate ships up to a distance of several miles. A similar system was installed in night fighters with some degree of success. Greater resolution and range had to wait until the development of high power microwave sources operating at much higher frequencies. This early work was done at Bawdsey Manor by a small team under the leadership of Dr E.B. Bowen.

Dr Bowen's team was small but the historical importance of the first detection of an aircraft by airborne radar must rank in importance alongside the original development of ground radar. Keith Wood, one of Dr Bowen's team, recalls this historic occasion.

Dr Bowen got a friend in the Air Ministry to alert him to plans for a RN fleet exercise in the North sea involving Coastal Command aircraft. We planned to have Anson 6260 [an Avro Anson fitted with 1.5 metre ASV] on top line for this opportunity, and at dawn on 4 September 1937 we took off from Martlesham Heath at 05.15 and headed out to the North Sea. The visibility was very bad and we couldn't see across the airfield. We were Dr Bowen, Sgt Pilot Naish and myself. We were fortunate to have Naish with us since not only was he an excellent pilot but also an ex-Merchant Navy officer and an expert navigator. Visibility was fine above 500 feet but below that it was very thick ...

After about 40 minutes flying at 10,000 feet we began to see several small echoes coming in to the maximum range (10 miles) of the time base. Flying towards these, to our great joy, into the end of the display time base at 10 miles came a huge echo many times bigger than we had ever seen before. One was from the *Courageous* aircraft carrier and to complete the picture, a number of smaller echoes indicating her destroyer escorts.

However, this was not the only outcome of the momentous flight. Suddenly several small but very distinct echoes appeared between the original transmitted pulse and the first and main pulse return from the sea directly below us. Moment laters Sgt Naish reported several swordfish aircraft had come up through the fog layer and were making up towards us. They had been dispatched to see what lone aircraft was flying around their ships. This then, was the first real air to air pulse radar contact, and under almost true operational flying conditions.

Robert Watson-Watt

Robert Watson-Watt was born on 13 April 1892 in Angus, Scotland. He studied at University College, Dundee, graduating with a BSc in engineering in 1912. In 1915 Watson-Watt joined the Air Ministry Meteorological Office (the modern Met Office), after finding that there were no suitable positions available at the War Office, which had been his first choice. The former were interested in the application of radio to detect thunderstorms, which Watson-Watt thought would have a practical application for pilots.

He found that while he was able to detect thunderstorms at a large range (a radio signal is given off as lightning ionises the surrounding air) the difficulty came in locating it, which was necessary in order to pinpoint the direction of the storm. He solved this problem by using a directional antenna, which when turned would show an increase or lessening of the radio signal, showing directionality. He then pioneered the use of the cathode-ray oscilloscope as a means of displaying the signal, and the system – a significant part of the later radar system – was in use by 1923.

During the 1930s as European countries began to re-arm, there was growing concern over the possibility of aerial bombardment of civilian areas. British air defence was considered not up to the task of holding off enemy planes, not least because bombers could approach at altitudes which anti-aircraft guns could not reach, and a lack of any reliable early warning system meant that fighter planes would be unable to become airborne in time to intercept them. At the same time, rumours of a Nazi 'death-ray' that could destroy a town with radio waves incited the Director of Scientific Research at the Air Ministry, H.E. Wimperis, to ask Watson-Watt about building a British version. Watson-Watt soon proved that such a device was a scientific impossibility, but in a secret memo he noted that radio waves could have an application in detecting aircraft.

The memo was entitled *Detection and Location of Aircraft by Radio Methods* and encouraged the Air Ministry to assign funding to Watson-Watt, giving him the chance to prove his theory. On 26 February 1935, a demonstration was carried out in which two antennas 10 kilometres apart received a shortwave broadcast, which was successfully interrupted by a Handley Page Heyford bomber.

On 2 April Watson-Watt received the patent on a radio device for detecting and locating an aircraft. Later trials at Orfordness increased the detection range, and by the end of 1935 it was up to 100 kilometres.

Watson-Watt moved his research team to Bawdsey Manor, and pressure to produce workable detection devices increased as the political situation in Europe worsened. The team began tests on a fixed radar tower system – later Chain Home – which would be an early warning system for incoming bombers. A special system of reporting the signals had to be created to allow the observers to relay information to the aircraft personnel accurately and quickly, but the first three stations were ready by 1937. Soon a chain of fixed radar towers sprang up along the English coast, and 19 played a key role in the Battle of Britain. There were over 50 by the end of the Second World War.

Watson-Watt also pioneered the invention of fighter-carried radar, as he realised that the Chain Home system had limited accuracy. He assigned Edward Bowen to produce a prototype no more than 90kg in weight and which required no more than 500 watts of electricity. The resulting Airborne Interception system was in widespread use by 1940, and played a vital role in ending the Blitz of 1941. It was later adapted for use on board ships, and thus reduced the threat from submarines.

In 1939 Watson-Watt became the Scientific Adviser on Telecommunications to the Air Ministry, and he travelled to the US to advise on air defence. He was knighted in 1942 in recognition of his contribution to the war effort. Ten years after this award he was given £50,000 by the British government for his work on radar. In his later years he had a practice as a consulting engineer, and lived in both Canada and the US, where he published his book *Three Steps to Victory* in 1958. He died on 5 December 1973.

CHAPTER II

Swanage

The basic techniques of Radio Direction Finding (RDF) could be extended to other areas, in particular to navigational aids. Early bombing raids on German industrial targets had shown great inaccuracies. Sometimes the target was not hit at all. Improved new aids were essential.

R.J. Dippy, one of the original scientists from Bawdsey, had suggested a radio system based on alignment of short duration pulses received from three ground stations to give a fix on a relevant map reference, but such was the urgency to get RDF up and running it had low priority. There was also opposition from the higher ranks of Bomber Command who had been brought up with the traditional method of navigation by astro observations and dead reckoning. This they believed was adequate, and unlike Fighter Command, who welcomed technological aids, wanted no fancy gadgets fitted to their aircraft. It needed the intervention of Churchill to resolve this reactionary attitude.

It was to R.J. Dippy's research team that I was assigned as laboratory assistant at the age of sixteen in early 1940. It was located in a former school in Langton Matravers, an uphill cycle ride from Swanage, where TRE had moved in 1939.

Mr Dippy introduced me to the scientific officer for whom I would be working. The other lab assistant in the laboratory introduced himself to me, informing me that as I was now the new junior it would be my job to brew the tea for the whole lab, including Mr Dippy, at 10.30am in the morning and 4pm in the afternoon each day.

I was assigned to assemble one of the sub units for the proto-type navigational aid, codenamed GEE. GEE was a hyperbolic navigation system, similar to the the current-day LORAN-C. The equipment was to be mounted in a wooden frame for fitting into a Whitley aircraft for flight evaluation. There was a great sense of urgency and no detailed drawings. I had to lay out, fit and wire one of the units direct from a freehand sketch. Equipment in those days was based on thermionic valve technology. The valves were plugged into bases, which in turn were mounted onto an aluminium chassis, then hard wired to their associated components.

R.J. Dippy

R.J. Dippy started work on radar development at Bawdsey Manor in July 1936. His proposals for a radio navigational system were not immediately used but were developed urgently in 1942 to improve bombing accuracy. The key advantage of Dippy's work was perhaps how successful his system was in returning bombers back to their aerodromes – it surely saved many aircraft and crews. In the US, with Dippy as adviser, the system developed into Loran-A. His systems were used on D-Day to provide precise navigation for the massive invasion fleets and airborne forces, and in the 'feint' invasion that fooled the enemy about the actual location of the attack. After the war he worked in New Zealand on electronics research and then he went to Australia where he became a divisional head of research at the Weapons Research Establishment near Adelaide.

Whilst I coped with the technical work reasonably well, my tea-making activities were not so successful. The first day Dippy came out of his office looking anything but pleased.

'Who made this tea?' he asked.

'I did sir,' I responded.

'It's too weak.'

'I'll make it stronger next time, sir.'

I eventually passed the tea test.

We were situated on the first floor of the building. Next door was an apple orchard. This prompted a joint research effort by the two lab assistants. It resulted in a long pole with a cocoa tin and razor blade at the end. Our neighbour must have noticed, for a message was received that the members of the establishment were welcome to as many of his apples as they wished.

One of the duties of the lab assistants was to collect components from the stores in Worth Matravers. This was a welcome break, particularly in the summer. At the bottom of the cliffs at Worth Matravers was an inlet from the sea forming a small isolated pool. We would clamber down the cliff, strip and cool off.

The flight trials of GEE illustrated that an aircraft could accurately locate its position up to a range of 200 miles or so, a distance sufficient to cover the German industrial complex in the Ruhr. The pressure of its development moved to the firm, probably A.C. Cossor, to engineer it for production.

I was transferred to the trainer group under G.W.A. Dummer. (We were all referred to by our surnames back then. I don't know Dummer's first name to this day.) Dummer's background must have been in heavy electrical engineering. He was older than most of the other scientists and may have previously been in a senior management position. I had the impression that I was more familiar with valve technology than he was.

Unlike the general thrust of the work at TRE, which was pushing into new techniques such as the evolution of microwave technology, essential for improved range resolution for night fighter systems, the trainer group took over where others had already blazed the trail.

This suited me, as I could be at a comparatively higher level in a lower level environment. Its purpose was to supply equipment that would allow recruits to learn operating procedures on radar systems without disrupting operation of the real thing. The relative failure to interdict enemy night bombers in the winter of 1940-41 could be put down in part to the speed with which the GCI and AI sets were thrown into the battle. Parts of the coastal chain had been in use for years against day bombers and personnel had had time to learn and adjust – but time was a luxury that no one could afford against the night raiders. The RAF had to be speedily trained but equipment was in chronically short supply. So training devices had to be designed and built. To do this it was necessary to generate pulse producing and timing circuits having such names as phantastrons, transitrons, multivibrators and flip-flops. Whilst working on these circuits Dummer would come up beside me nervously jingling the change in his pocket saying, 'Well

Goult, how is it going?' I might reply, 'Its going well sir.' In which case, he would take an interest while I demonstrated the circuit. Or I might reply, 'I'm in a little difficulty sir. It isn't going as it should.'

'Keep at it, keep at it,' he would reply, looking anxiously at his watch, and hurrying off.

I was happy to be left to my own devices and there were plenty of experts around willing to give advice if necessary. Thus Dummer's team produced an AI trainer that simulated the appearance of the AI indications on a radar screen so that a night-fighter crew could practise homing in on a target without leaving the ground. How many thousands of hours of flying time and gallons of fuel were saved by the use of the simulators by the end of the war is difficult to judge, but the numbers are large. In the early months of enemy night bombing their losses were negligible but they rose steadily as the training bore fruit, so that in May 1941 102 night bombers were shot down and 172 were probably destroyed or damaged. Enemy losses grew from less than half a per cent to more than 7 per cent and the attacks more or less ceased. This tangible victory was won by the skills and bravery of the fighter pilots, plus a few GCI sets and perhaps 100 AI sets.

On my 17th birthday I joined the Home Guard. There were no Dads in our Dad's Army platoon. The majority were young university graduates. Captain Scott, the platoon commander, bore no resemblance to Captain Mainwaring. He had previously been a regimental sergeant major in the Grenadier Guards!

Within a few weeks of my birthday France fell and the remnants of the British Army in France were miraculously plucked off the Dunkirk beaches by the Royal Navy and a flotilla of small boats. The army had to be regrouped. The coastal defence of Swanage and TRE was left to the Home Guard. I had only fired my rifle once! We had to come to work in uniform complete with a band of ammunition and our rifle ready to hand. We were the defence force on this part of the coast. Hitler missed an opportunity. There was only one major road out of Swanage. A parachute drop on that road could have resulted in the capture of the cream of British scientists.

Unknown to me at the time I was working among young scientists who were later to achieve distinction in other areas. Dr Lovell, later to develop the Jodrell Bank Radio Telescope probing the extremities of space, was developing airborne radar at the Langton Matravers site. The school set up at TRE to teach the new radio technology, where I studied in the evening to obtain the first of my qualifications, was headed by J.A. Ratcliffe, seconded from the Cavendish lab in Cambridge. Another junior scientific officer later became Astronomer Royal.

The occasional bombs that fell on and around Swanage were probably the result of individual German aircraft unloading after missing their targets at Portsmouth or Southampton. The laboratory storekeeper was lost in one such raid. The site at Worth Matravers was strafed with machine-gun fire on a couple of occasions, but this was hardly the response that would have been expected if the true nature of the work was realised. As with the code breaking activity at Bletchley Park, the location of radar development was never discovered. Unlike Bletchley Park our existence was well known. There was an ongoing jamming and anti-jamming 'Radar War'. Our response in this activity was generally faster than the Germans'. TRE was staffed mainly by academics who were not averse to 'lashing up' equipment, whereas the Germans' with their more formal approach were reluctant to move from 'good engineering practice'.

Danger or no danger, the powers that be decided, probably with some justification, that TRE was to move to Malvern College in Great Malvern. There was some communication to the senior figures concerning the possibility of an attack by German paratroopers on TRE, but A.P. Rowe thought this was ruse by London to get the establishment moved away from the coast out of the way of the eventual, inevitable invasion of Europe. A hundred Pickford removal vans were assembled to carry out the move. I was to ride shotgun in one of the vans. Who they expected to attack us on the road from Swanage to Great Malvern I cannot imagine. In some ways the glory days of TRE had been at Swanage. The equipment needed to defeat the night bombers had been developed there and GEE been introduced to guide Allied bombers across Europe. Oboe and H_2S were developed at Swanage and the benefits of centimetre radar were being felt by both the Army and the Navy. Coastal Command was evaluating ASV, a Swanage product.

The removal of the equipment and those that went with it was very efficiently organised. This was not the case with the removal of people other than those being transported with the material. In general it was expected that those with cars should fill them to capacity. Those who couldn't get lifts were to be given rail vouchers. In typical British fashion it sorted itself out in the end.

None of this would have affected me had I not been developing a blossoming friendship with a girl from the print room. It was decided that rather than recruit new support staff in Malvern it would be less time consuming to retain the existing staff. Beryl, my new friend, aged sixteen, would be travelling to Malvern. She asked me to travel with her, but I explained that duty called me to defend the convoy from attack by the unseen enemy. However, I approached a senior scientist I had never spoken to before and have never since, persuading him to take her in his Lagonda, and promised to meet her at the billeting office in Malvern College.

CHAPTER III

Great Malvern

I met up with Beryl as soon as I arrived at the college. She had been allocated a room at Malvern Link. The office map showed this separated from Great Malvern by a common, which we made our way across and located her billet. I left her haversack at the door after she had met the house owner, and started to make my way back across the common when I heard plaintive cries of 'Ian, Ian' behind me. A tearful Beryl was rushing from the house.

'It's awful,' she said. 'There is nothing in the room except a bedstead.'

We made the trek back to the billeting office. They had had a call from the house saying that Beryl had rushed out before they could explain that they had not had time to make the bed and furnish the room. Eventually she was persuaded to return.

'That means another route march across the common,' I remarked ruefully.

'Oh no,' said the WVS lady in the office. 'There is a bus stop outside. A penny bus ride will take you right to the door.'

That evening we went to see a film called *Hellzapoppin'*. It could not have been more appropriate. The cinema was filled with TRE personnel. The film was so ridiculous, mirroring our situation, everyone just fell about with uncontrollable laughter.

The locals were not laughing. Our invasion was not well received. Who were these people upsetting the tone of the place with their corduroy trousers and tweed jackets patched at the elbows and cuffs? Furthermore, any spare accommodation had to be given up for compulsory billets at five shillings a week. There were stories of eminent scientists being offered beds in garages.

Malvern seemed to have existed in something of a time warp. There was a high-class haberdasher in the main shopping centre that up to the outbreak of war had rooms in the attic for live-in shop assistants. It is a lovely town, terraced along the east side of the northern end of the Malvern Hills. Fine Victorian houses were at the upper levels – ordinary Victorian houses at the bottom level. The war had mostly passed the town by. A bomb had apparently fallen a few miles outside the town but no one knew where. This invasion by a thousand outsiders was distinctly unwelcome.

I was billeted in the house of a retired headmaster. I think he was trying to be friendly when he suddenly asked without preamble 'Have you read your Trollope?' I admitted I hadn't. This must have put me beyond the pale. Later, after I had left, I learned that his son had been killed in Burma.

Billetees were not given food at the billet. The Winter Gardens Concert Hall was taken over as a canteen, staffed in the first instance by the WVS, and meant I had a half-hour walk to breakfast. On my first morning I was surprised to see J.A. Ratcliffe of the Cavendish labs serving the porridge.

'You must have been up early,' I ventured.

'Oh, I climbed to the top of the Worcestershire Beacon before I came here,' he replied.

In the daily food queues at the Winter Gardens you never knew who you might be next to. In general for lunch, the whole of a particular laboratory would go together. This created a good spirit of camaraderie throughout the establishment. It also tended to create a community within a community.

The Winter Gardens was restored to its former use after a new canteen was built in the Malvern College grounds. This helped to appease the local community, and there was a further improvement of attitudes and a fading of prejudices after Air Marshal Sir Phillip Joubert – well known for his radio briefings on the state of the war – addressed the local council and Chamber of Commerce. Sir Phillip had written a short film about the Spitfire in the Battle of Britain, rushed out in 1940.

Coastal Command began the war at a disadvantaged position in terms of its capabilities for defending shipping in what became known as the 'Mid-Atlantic air gap'. Pitiful interwar antisubmarine training together with the domination of offensive strategic bombing in the RAF meant that the maritime air arm continually struggled for resources for the U-boat war. In June 1941, Joubert would become air officer commanding-in-chief, Coastal Command and try to get more aircraft assigned to the defensive role. Joubert was at odds with Bomber Command and the Air Ministry, who were intent on keeping the offensive bombing strategy as the number one priority – Joubert also had to argue against Bomber Command's key supporter, Churchill.

Before long I moved to Mount Pleasant, a hotel that had been taken over as a staff hostel. There must have been nearly a hundred of us there, which was in the upper part of Malvern. The Worcestershire Beacon was behind the hotel, and the dining and living rooms looked over the Severn Valley, which at that time, but regretfully no more, was full of plum blossom in the spring. With this move, life could be lived exclusively within the TRE 'estate'.

In the meantime work went on apace. The speed at which the whole establishment was up and running again was remarkable. Everybody and anybody did everything and anything that needed to be done. The speed of technical development was also remarkable. More advances were made in a few years than had been in the previous twenty. It changed the way in which war was waged. It also set the scene for post-war electronics development.

One of the most significant changes was made manifest during the Battle of the Atlantic. The German U-boats were operating with impunity to devastating effect, sinking 100,000 tons of shipping in 1942. The tables were turned with the development of microwave technology.

The coastal chain stations operated at a frequency of around 45MHz – a wavelength of 6.6m. This was not good enough for airborne or ships' radar. Shorter wavelengths were required for the resolution required to pinpoint accurately the range and bearing of submarines. Some success had been achieved initially with lower wavelengths, typically down to 1.5m. However, the U-boats were fitted with receivers able to detect transmissions at this frequency giving them time to dive before the aircraft had the necessary visual range for a successful attack.

A reduction to a wavelength of 10cm was required to obtain better resolution and achieve surprise against German U-boats surfacing to recharge their batteries. Furthermore, it would be possible to home in on the periscope of a submerged submarine. Low power microwave sources already existed. However, to obtain a useful range for realistic searches over the vast area of the Atlantic Ocean a more powerful source was needed.

The breakthrough was a result of close and enthusiastic cooperation between TRE, universities and industry. Birmingham University was tasked to develop an improved microwave source. It is sometimes possible to upgrade an existing device, but the klystron tube power was limited by the need to focus the electron flow into a very narrow beam. Two of the research team – Howard Boot and John Randal – worked out a way to link an earlier development that had showed no promise (the magnetron) to the klystron to produce a circulating electron beam coupled to a series of resonant cavities. Their handmade model produced power an order greater than the klystron. Further development by GEC boosted the peak pulse power output to 15 kilowatts. This was the cavity magnetron, the same device as is used today to cook the Sunday joint.

Despite the uprooting of TRE from Swanage to Malvern, the new microwave source was rapidly engineered in conjunction with GEC into the Air to Surface Vessel (ASV) radars. Not only did new technical problems have to be resolved, but political prejudice and other road blocks had to be overcome – not least of which was the loss of key members of the research team in an aircraft crash while carrying out flight trials with the only existing prototype of a ground mapping radar (codenamed H_2S, 'Home Sweet Home'). It was at this time that the cat's whisker, virtually the same as I had used when playing with my first radios at the age of twelve, had a new lease of life.

'Home Sweet Home'

In late October 1941 A.P. Rowe held a 'Sunday Soviet' – a brainstorming session – to think of ways to help Bomber Command to bomb blind. There was no immediate break-through but following work by P.I. Dee on the application of centimetre wavelengths, it was realised that a rotating cen-timetre beam scan intended for locating ships at sea could also locate conurbations – towns instead of ships. A flight was made to Southampton with a modified centimetre A.I. scanning the area ahead of the aircraft, Sure enough, the city showed up strongly on the cathode ray screen. During another flight above cloud across Midland towns, photo-graphs were taken of the screen indications of built-up areas. When the still-wet prints were laid on Rowe's table, he com-mented 'This is the turning point of the war.' H_2S was born. The name was suggested by Lord Cherwell, referring to the ability to zero in on a target – 'Home Sweet Home'.

Thermionic valve technology as used for conventional radio reception could not be used at microwave frequen-cies. The wavelength of microwave signals was shorter than the physical size of conventional radio receiver diode detec-tors. This problem was resolved by Dr Herbert Skinner at TRE who encapsulated a small piece of silicon crystal with a small length of spring wire pressed against it into a con-tainer not much bigger than a pea. This was the first silicon chip, although no one realised it at the time. Dr Skinner was happy enough that it performed the required job of detecting low-level signals in microwave radar receivers. It was left to scientists at the Bell Labs in the US to unlock the secrets of semi-conduction in silicon and pave the way for the silicon chip era that followed. In the 1940s we led the world in the development of microwave technology. It was not to last. I was aware of a stream of American visitors passing through our laboratory, viewing the AI (Aircraft Interception) simulator, and other equipment. (It can be argued that radar and its technological derivatives was one of the key breakthroughs the British gave up to the Americans post-war 'for free', together with the jet engine.)

The breakthrough in microwave technology was great – up to a point. Its implementation in anti-submarine warfare in the Atlantic required production and deployment on both sides of the ocean. With this in mind Churchill and Roosevelt came to an agreement even before the Americans joined the war that all our technologies should be pooled in exchange for much needed aid. Thus the secret of the new microwave transmitter source and all our associated RDF technology was passed to the Americans – and the Battle of the Atlantic was won. During the late spring and summer of 1943 Coastal Command kept TRE abreast of events as the U-boat menace was defeated by centimetre ASV. The U-boats were forced to surface in daylight and even resorted to staying on the surface in threes and fours and actually engaging with attacking aircraft. In one month, more submarines than merchant ships were sunk. From June onwards merchant shipping losses were negligible.

On entering the war the British Branch Radiation Laboratories (BBRL) and (ABL), staffed by American scientists and service technical officers were set up adjacent to the TRE laboratories at Malvern College. We were able to produce magnetrons in their hundreds; American industry made thousands, and millions of cat's whisker diodes. They renamed RDF

(Radio Direction Finding) 'RADAR' – **Ra**dio **D**etection **A**nd **R**anging – and post-war the massive American radio industry took the lead in radar development.

Another technology that had worldwide application was the Plan Position Indicator (PPI), its purpose Ground Control Interception (GCI), directing a night fighter within range of an enemy aircraft. Today this same display is used for the control of civil aircraft movements worldwide.

Development of the PPI display was another facet of Dummer's group. This was carried out at the site at Worth Matravers by Dr Franklin and Mr Florida (no first names again) during the time I was at the Swanage location.

I needed access to the display unit to establish where to feed dummy pulses for the operator-training unit. I recall that Florida would reach up to the antenna connector of the transmitter and sparks would fly off the tips of his fingers to the connector. Low repetition rate pulses at high frequency are not lethal. Not so the DC voltages of several thousand volts in the vicinity of the cathode ray tube display. It was in the display unit that I had to probe to establish where to feed my dummy pulses. One just had to be careful. When using valve technology there were always high voltages around.

I noticed that quite a few of the senior scientists started to disappear. No one mentioned where they had gone. I later found out that some had joined the atomic bomb programme and others a guided weapons programme. To make good the gap left by those that had disappeared they must have recruited physics teachers from schools, as who should I meet but my former physics teacher working in another lab. He had been teaching me only about one year previously. More surprising than meeting him was a request to look at some notes on the work I was doing. I wondered if they would come back covered in red ink. They were merely returned with a polite thank you.

Fresh young university graduates were also recruited, not only to fill the gaps, but also to push forward further developments. To complement these expanding operations, selected girls leaving school with good exam results were recruited as lab assistants. They were billeted at Parkfield, their own hostel, within walking distance of Mount Pleasant. The 'estate' became full of young people.

In October 1943 a train pulled into Great Malvern station and two strangers alighted, Graham Rushton from Cambridge, and Laurie Hinton from Imperial College London. They discovered they had a common destination and together made their way to Malvern College.

After registration they were taken to their billets. As they were driven round Malvern in a large comfortable staff car they felt that their destination would be a comfortable army mess. They were mistaken. They were invited to alight in the disused Malvern Wells Station goods yard and shown two brick huts six feet from the main line to Hereford. Each occupant had a bed and a wooden box for his clothes. They would have to walk for all meals across the common. By late 1943 there were no more billets in houses available. Graham and Laurie later became good friends of mine.

Laurie went to work on microwave technology in the centimetre receiver group. Graham joined Dummer's group. Inevitably I met up with both of them and other young graduates. Various informal groupings were established and one circulated between them. We generally congregated in the evening at the Piers Plowman Club, open to all members of TRE, but frequented mainly by the younger members, including the new intake of lab assistants.

Dances were held on Tuesday, Friday and Saturday nights at the Winter Gardens. There was a resident dance band and we went to most of the gatherings. Our dance partners soon became our girlfriends and our names became coupled. The question would be asked, 'What are you and Joyce doing next weekend?' The pairings were not binding and frequently changed.

Weekends were, however, far from free. We worked according to the pressures of the job. It was later recommended that seven-day working should be avoided to prevent stress. A five-and-a-half day week was the norm. Stress was unlikely in our circle. We would meet up in the evenings and weekends when we could. Work was not discussed. In fact we were under instructions not to discuss work outside Malvern College. This was the antidote.

Sex was not on the agenda, at least not in our particular circle, especially since the girls' hostel required them to be in by eleven. Naturally we would always 'see them safely home' to the door – or to the drive. It got fairly crowded at eleven pm! Nor was there much drinking. The pubs were usually crowded with service personnel, limiting our access, and the Winter Gardens was 'dry.'

The main mode of transport was by foot or bicycle. Cars were a rarity, even amongst senior staff. Pay for even the most senior scientific officers was pretty meagre. Travel in Malvern was mainly steeply up or down, so even though everyone had a bicycle they were seldom used locally. It was at weekends that they came into their own – for youth hostelling in the Wye Valley. It was Laurie Hinton who organised most of these weekend outings. They cemented the comradeship of our informal circle. We had the roads to ourselves. Simmons Yat and Tintern Abbey were deserted.

I acquired a car. This was a leftover from the requisitioned garage in Felixstowe. My father said that rather than let it rot I might as well have the use of it. This opened up new and varied opportunities. I was the only car owner in Dummer's group and in our social circle.

As previously described, Dummer's group was the design authority for the training equipments that were being manufactured and going into service in the RAF. I now had a roving commission (and extra petrol coupons) representing the design authority both at the manufacturers and at RAF locations.

I recall my first visit to RGD (Radio Gramophone Development company) in Bridgnorth. I was met by the technical director and taken to lunch in the boardroom. The afternoon I spent in his office going over the drawings and discussing the production test procedures.

In the evening I was looking forward to a quiet drink in the bar of the pub where I was staying for the night. This was not to be. I was accosted in the bar by a very offended ministry man, the resident Quality Control Inspector at RGD. I had broken established ministry protocol. I should have made contact with him immediately on my arrival at the firm. I apologised profusely. As it was obvious to him that I knew absolutely nothing of ministry protocol he spent the rest of the evening filling me in. He was quite happy however to let me check out the equipment on the following day and sign the release note.

Each visit was a different experience. A firm I visited in Wembley transported me back to the previous century. Had I met Charles Dickens I would not have been surprised, so Victorian did the set up appear. Overhead shafts ran the length of the production area. From the common drive shaft belt drives fed down to individual operating machines turning out washers, spring washers, hooks and eyes and all the items you might expect to see in the local hardware shop. It was for a different sort of hardware that I had drawings.

The drawings were of a mechanical 'crab' that moved about on a map following the flight path flown by a link trainer or similar simulator with cockpit controls. The pilot would be attempting to close on a simulated radar echo set up by the instructor. As can be imagined the crab mechanism was quite complex. I was not the design authority for this device, but nonetheless I represented the design authority at the Wembley manufacturer. In the boss's office were gathered the directors and production engineers. They had all been security cleared, so I was able to explain the purpose of the equipment. They, like most of the rest of the country, knew nothing of radar. According to the media the night fighter pilots had cat's eyes, as a result, they said, of eating lots carrots. In fact they had microwave radar. The engineers, most of whom were at least twice my age, were the designers of the machines that turned out the nuts, bolts and washers. They made a superb job of the crab.

The car affected my social life. Laurie Hinton had left his digs and taken up residence in Mount Pleasant. The sleeping accommodation was reminiscent of our school dormitories. There was an absolute minimum of furniture allowing three or four to a room. Laurie and I seldom saw our third occupant, as he was usually asleep by the time we got in. Life in the TRE estate was somewhat parochial. My car gave us freedom of movement and some Saturday evenings we would venture further afield. On one such outing we linked up with two ATS girls at a dance in Kidderminster. They were good friends, so we made a very pleasant foursome and met up a few times. Laurie and his girl drifted apart, but I continued to date

Eleanor. She wanted to learn to drive and so became the first of many pupils. Our lessons never lasted long. She was very affectionate. In the end Eleanor got posted and we lost touch.

I also visited various Bomber Command stations. The purpose was to commission and instruct on the operation of the 'Oboe' and other training equipment. Oboe was a cat and mouse affair. Two ground-based transmitters and receivers spaced several miles apart were referred to as the cat and mouse stations. Each would simultaneously transmit a short duration pulse. The time delay equivalent to the distance from each station to the target would be known. An aircraft would receive dots or dashes depending on his position relative to the cat. The signals would merge to give a continuous tone when at the correct distance. This enabled the pilot to fly along the circumference of the 'cat' circle towards the target. When he crossed the range defined by the mouse he received a signal defining his position over the target. The trainer enabled the pilot to practise the approach to the target without using aircraft. It was also used for operational research. Invariably the pilot would overshoot the cat distance, weaving back and forth across it. This was overcome by giving the continuous tone when the rate of approach was correct. Of course when 'on the beam' the required rate of approach would be zero.

When visiting Bomber Command stations I would be the only civilian in the officer's mess. This could be very lonely, so my car came into its own. The bases were usually in fairly isolated areas, particularly in East Anglia. Very few, if any, of the officers had their own transport. I could offer an evening out in the nearest town or village.

A variation of Oboe was G-H. This combined two existing navigational systems, enabling any number of aircraft to use the system simultaneously. I was involved in the development of the G-H trainer. Ted Ley, one of our social circle, was carrying out flight trials. This led to my first flight. Ted suggested that I join him to see the system in operation, to which I readily agreed.

Such was the informality of the establishment at that time, he simply rang up to say he had a flight organised and I immediately went over to his lab and we set off to Defford aerodrome, showed our passes, and drove to the appropriate hanger. There Ted introduced me to the pilot, an officer in his late teens – no older than us. We picked up parachutes and boarded the massive four-engined Lancaster bomber. We were to do a dummy run over East Anglia. On the way Ted sat in the co-pilot's seat and took over the controls. I observed the operational procedure, aligning and keeping aligned a pair of pulses displayed on a cathode ray tube.

On the way back Ted pointed out his uncle's farm below. 'Let's shoot it up,' exclaimed the pilot, banking the aircraft and putting it into a steep dive. I was standing on the wing spar looking out of the astrodome. When it looked as though we were going to hit the farm the pilot pulled out. I should have known better, the G force made my knees buckle and I went down on the wing spar with a thump. Fortunately Ted and the pilot were both looking forward at the time and did not notice my embarrassing tumble. I had backache for weeks.

On another occasion I disgraced myself during a different flight trial. I was again in a Lancaster observing BABS (the codename for an aircraft beam approach beacon system). We were approaching the airfield in thick cloud. We had been airborne for some considerable time carrying out associated trials, and I had been standing all the time. I decided it was time to sit down and did so, on the box behind me. It was the power unit for the landing cathode ray tube display, which promptly went blank. We were in thick cloud approaching the runway with no guidance. 'No problem,' said the pilot. 'If I maintain this course and rate of descent we should hit the runway.' We broke cloud over the undershoot. Today I would have 'received a rocket', but back then it was 'just one of those things'.

The failure rate of airborne radar equipment was very high. The important criterion was that it should last for the next operational mission. It would often be removed and checked before each mission, on an almost daily basis. Many of the radar officers and technicians I associated with around the air bases I visited were Canadian, trained in Canada to a very competent level. I was always confident that they would understand the operation and purpose of any new piece of training equipment I introduced.

The Germans were working on a secret weapon. This turned out to be the flying bomb and its successor the V2 rocket. One of their experimental rockets fell in neutral Swedish territory, and Churchill quickly arranged for a team of scientists to visit Sweden to examine it. How could this affect me? Imagine my surprise when Dummer came up to me and said almost as an aside that it was believed that the German rocket had a radio guidance system working at around 200MHz. He had been asked to develop some form of jamming system. Would I see what I could do? I could only imagine that all the jamming experts were busy on other things. In this, I learned later, I was correct. They were preparing radar decoys for the planned Operation *Overlord* invasion of France.

My effort was pretty crude and basic. I recall I worked at it over Christmas, either 1943 or '44. All I could do was to make up a basic 200MHz oscillator source and swing it back and forth over as wide a band as possible in the hopes of interfering with any guidance signal at or around the frequency. I heard no more of it. In fact although there was some electronic control in the V2 it was not radio controlled. Nobody bothered to tell me. I made a report and got on with other things.

I never made the cricket team at school. No matter. I played on the cricket square at Malvern College. I think we were

'Oswald'

Gwen Reading worked at Bawdsey on 'Oswald', code name for the equipment used in Operation Big Ben, the effort to locate and bomb the V2 rocket launch sites. She recalled: 'All our work on radar stations was considered very, very secret; but when "Oswald" was introduced we were told that, if possible, the work of tracing the V2s was even more secret. I would say that set was about 4 ft 6 in high, 2 ft 6 in wide and 2 ft deep. The screen was small, probably less than a foot square. We were only allowed to watch the screen for 15 minutes at a time. The V2 rocket showed as a curving, thread-like trace. On seeing it, the operator yelled "Big Ben ay Bawdsey" down the line to Stanmore.' The film of the trace was then hurriedly developed and analysed to pinpoint the launch sites. Gwen believed that a four-minute warning could have been given to London of an approaching V2, which was never done, but as she pointed out, 'I suppose it would have meant the entire population of London scurrying to shelters to avoid the explosion, which – large as it was – was only going to affect one area. There was no way of pinpointing where the V2 would land.' Gwen's story is taken from *Radar: A Wartime Miracle* by Colin Latham and Anne Stobbs.

playing our sister establishment, an army radar research group at Malvern Link. I recall with some satisfaction that I was amongst the highest scorers at 22 not out. This was I believe an isolated use of the college cricket ground, which was maintained with proper care and respect.

There was, however, a regular hockey team. We usually played a men's match on Saturday afternoon and a mixed match with the ladies team on Sunday. Needless to say I teamed up with one of the ladies for dinner at the Westminster Hotel in Colwell, about an hour's walk over the Worcestershire Beacon. No problem in those days.

The Westminster Hotel has a place in the radar story. The local pubs, although comparatively respectable, were seldom frequented by the establishment. If Dummer's group – of which I was a member– had an outing or celebration of any sort it would invariably be at the Westminster. The group was friendly and there was little class distinction or racial bias. One member of the group, Chander Nath, had been stranded at Bangor University at the outbreak of war. Unable to return to India after completing his degree he had ended up at TRE. He was a cheerful and popular member of the group, which consisted of around 16 to 20 members and a floating population of service personnel. It was predominantly, but not entirely, male.

One such outing to the Westminster was arranged by Guerney Sutton, a senior scientist in Dummer's group. There was no particular reason for the group outing, other than meeting up away from the workplace. The work ethic in the group was extremely high – as it was in the whole country after Dunkirk – and nowhere higher than in the person of Guerney Sutton, who chain-smoked and often ignored the lunch break as an unwelcome intrusion. He was working, I think, on the AI (night fighter interception) trainer. He was personally easy going and popular. There were about a dozen of us meeting up and we were enjoying a pre-dinner drink in the bar, when Guerney Sutton walked in and announced that there was a change of plan and we were all going to his place for beer and sandwiches. No reason was given, but as far as I was concerned if Guerney Sutton said it, then it was OK by me. We had a pleasant and convivial evening.

It was later that I learned that the manager had taken Guerney aside and explained with regret that they did not serve coloured people. I am glad we walked out. I did not go to the Westminster again until my 21st birthday, but that's another story.

Living in a closed community guarded by military police could create personal and family problems. Some of the living accommodation was within the college grounds. I was about to enter the establishment one day when two anxious parents desperate to get in touch with their daughter approached me. She had informed them that she intended to marry the head of the photographic department. Did I know their daughter? I did not. However, I said that although I did not know the head of photography personally, he was highly respected. This was true. This seemed to relieve the parents. As far as I recall the story had a happy ending.

I very nearly got engaged myself – twice. I fell desperately in love with a girl from the typing pool. Jean was a local girl. After working hours we were inseparable. Any activity automatically included both of us. She was accepted, or perhaps I should say tolerated, by the rest of the girls from the Parkfield hostel who formed the rest of the female part of our circle.

She was the eldest of four, the youngest of which was still a baby. Their father was away in the army. At the weekends she would sometimes relieve her mother. We would then take the whole family, including baby, out for the day. I became a part-time father.

I taught Jean to drive. By today's standards the car, a 1934 Hillman Minx, would not be considered fit for the road. It had leading shoe brakes and unless perfectly adjusted they would grab and lock. At speeds above 45 mph sharp application of the brake could lock the offside front wheel and slew the car over to the opposite side of the road. I could anticipate this and drive accordingly. As far as Jean was concerned the accelerator moved the car forward and the brakes were for stopping. Why complicate the issue?

Worse was the condition of the tyres. You were likely to be stopped if you had a tyre with a decent tread to explain how you got it! This meant running the tyres till the canvas showed. I carried a puncture repair outfit and tyre levers, and often had to make repairs by the side of the road. In desperation I acquired two motorcycle tyres having the right diameter but the wrong cross section. I inflated the inner tube until it moved the tyre out to its correct position against the rim of the wheel. They gave no trouble.

It was the car that initiated the first of many rows between Jean and I. She would corner too fast or drive out of a side road into a main road without stopping or looking. We only survived due to the scarcity of other traffic. As we did survive, she resented being corrected.

We had been linked for two years, but there were strains appearing in the relationship. The rows left me emotionally drained, but were usually quickly made up. The crunch came on VE day, which also happened to be my 21st birthday. I had booked for a dinner-dance at the Westminster Hotel. It was nearly three years since the walkout mentioned earlier, and time is a great healer. It should never have been thirteen at the table, but I could not see how I could exclude any of my friends of the past four years.

Laurie and I had agreed to catch the train to London on the following day to join in the Victory celebrations. Jean had decided that I should remain in Malvern. She had already begun to decide what I should and shouldn't do. This was not on. We were barely on speaking terms. The whole group was aware that our relationship was cracking up. Despite this we had an enjoyable evening. On the following day Laurie and I went to London. Jean soon found a new partner.

With victory in Europe the establishment began to run down. Dummer had already moved on to investigate and improve the packaging of radio equipment. The emphasis was on the continuing war against Japan in the Far East. Equipment that worked perfectly well in Europe would break down in the tropics, often being dud on arrival. Many components relied on a wax coating for protection, which simply melted in the heat. Also, the manner in which they were packed for shipment was inadequate. New regulations and specifications were established.

In addition, Dummer initiated a programme of integrated circuit technology, but it was ahead of its time. Integrated circuits only became a viable proposition for silicon chip technology many years later.

Mr Dummer's deputy ran the group for a short period. He then moved on and by default I found myself the senior member of the trainer group and therefore effective head of department. I had neither the rank nor the qualifications for the position, which should have been held by a technical officer. Nevertheless I found myself in charge. The work was mainly a tidying-up process and further liaison with the RAF, particularly with some remaining installations of the Oboe trainer. The atmosphere at the RAF bases had changed. It was lethargic to say the least. The adrenalin had ceased to flow. Most were conscripts and now their *raison d'etre* had gone their only interest was how long before demobilisation. This atmosphere pervaded TRE to some extent.

I had a reputation among the lab assistants (I learned later) of being hard and strict. To their surprise I now allowed them to 'fly' the link trainer, to write home to their parents, and to sit and chat in the tea area. They were exclusively female. It was a reserved occupation, but they were now being released to return home. This included my current girlfriend. It was all very depressing.

I had acquired an interest in flying. The war against Japan looked set to go on for some time. I went to the naval recruiting centre in Worcester with a view to joining the Fleet Air Arm. 'No way,' they said. A few days later I knew why. Atom bombs were dropped on Hiroshima and Nagasaki. That ended the Japanese war.

Eventually the trainer lab was closed down. The college was being prepared for the return of the school.

With the end of hostilities the Americans moved out of the laboratories adjacent to the college. They were taken over by the instrument department of the Atomic Energy Research Establishment that was to be set up at Harwell. I was transferred to this department and would eventually end up at Harwell. This was not to my liking as I had no wish to go to Harwell or to work at an atomic energy establishment. Although the work on instrumentation would be of interest, the main thrust of the work would be atomic physics, of which I knew nothing.

My parents were about to set up business again in Felixtowe. Whilst doing this they had no income, so my sister and I sent part of ours to keep them going. They needed all the help they could get. I was no longer in a reserved occupation and decided to give in my notice and join my father.

This was more than a door closing. It was the end of a life – but it was not the end of radar and me.

PART II

Continuing Development
in Industry

CHAPTER IV

TACAN – Tactical Air Navigation

I soon realised that the family business in Felixstowe was not for me. I yearned to be back in research and development work. Government establishments were no longer recruiting, but reducing commitments due to post-war austerity. I was lucky that an ex-TRE colleague suggested that I should look to industrial research, referring me to another ex-TRE colleague at Standard Telephones and Cables, who were working on some interesting radar projects.

It was not so much an interview I had with my future group leader, who was aware of my work at TRE, but rather a guided tour – a tour of his development labs together with an outline of the projects he had in hand. He was universally known as 'Cap', short for Capelli. This name extended beyond STC, for he was well known throughout the industry and in the scientific branches of the Ministry of Defence.

Ex-TRE staff that had not returned to academic work at the universities from where they had been recruited were spread throughout the radio industry, which, due to their influence, was developing into the new electronics industry.

STC was one such firm. Capelli's group was one of many in the Radio Division at New Southgate. The division supplied transmitters for the BBC and Royal Navy. It was also in the forefront of airborne radio communication equipment to the fast-growing aircraft industry. Capelli's group was developing airborne navigational equipment based on the radar technologies developed at TRE.

TACAN (Tactical Air Navigation) was one such navigational aid, being developed by our parent company in the US under contract to the Defence Department. The military TACAN navigation system was an evolution of radio transponder navigation systems that could be traced back to the British Oboe system. It used a combination of pulse technology for range and analogue ground-based spatial signals for bearing. A ship or aircraft could obtain range and bearing relative to a known ground-based (or ship-borne) transmitter. The aircraft initiated the pulse transmissions to which the transmitter responded. This resulted in very complex airborne equipment using nearly 100 thermionic valves and transmitting high power pulses at the lower end of the microwave band.

STC's brief was to engineer an anglicised version of the equipment, for manufacture in the UK. There was no correlation between British and American valves and components at that time. This was to eventually change after the setting up of NATO.

I was assigned to John Birchenough, also ex-TRE, as a junior engineer. Birchenough had worked on the new microwave techniques at TRE and was to be responsible for the development of the high power transmitter pulses. At the time that I joined the group the project was still in the initial stages. It was necessary for one of the British team to go to the US to familiarise themselves with the detailed aspects of the equipment. For the airborne equipment this task fell to Birchenough. He was gone for about two months. This included the Atlantic crossings by sea, adding about eight days to the trip. He outlined some of the specialised test equipment he would need which I was to develop in his absence. Once again I was working on my own.

It was only after I left STC some twenty years later that I appreciated the back-up facilities that were available to the development teams. Each lab with three or four engineers would have its own lab mechanic. There was also a model shop dedicated to making up specialised equipment and assembling fully engineered prototype units. Draughtsmen came under the administration of the chief draughtsman but individual draughtsmen were allocated to work for particular projects and worked closely with the project engineers. All tasks could be carried out on site. There was a chemistry lab, casting and toolmaker's shop and wood shop, amongst others. The wood shop might sound an anomaly in a communications industry until it is realised that private branch telephone exchanges were manufactured on site, and they were housed in wooden cabinets. The wood shop also made the lab benches and other furniture used on site. These facilities meant that the engineers could concentrate on pure development work.

When Birchenough returned I had completed the work he had left me and we set about developing the radio frequency part of the airborne transmitter. He dealt exclusively with the final power amplifiers. Under his guidance I developed the oscillator source and frequency multiplier chain to feed the power amplifiers.

This had an interesting knock-on effect on my future. I was undecided whether to use a pentode valve for the second multiplier going from 120MHz to 360MHz, or a grounded grid triode. I wrote a technical paper, which came out in favour of a grounded grid triode. Such a paper would have a limited internal distribution and undoubtedly be read by the chief engineer, one C.E. Strong.

C.E. Strong was a contemporary of Marconi, and, in fact, together with the parent company in the US, set up the first commercial transatlantic radio communication link ahead of Marconi. Unknown to me, during his work on early radio communication he had taken out a patent on the grounded grid triode. Did my choice of grounded grid triode affect his attitude? Strong had set up a Standard Radio Engineering Society, a forum for the reading of technical papers, in particular for the younger engineers. He initiated a yearly award, for which I presented a paper on the special stable oscillator I had developed as the transmitter source for airborne TACAN. It was not particularly innovative, but I won the award for that year.

Some time later, after I was no longer on the TACAN project, a colleague pointed out that my paper on the grounded grid triode and its comparison with the pentode was flawed. The performance of the grounded grid triode was as predicted. It was the use of a similar method of prediction for the pentode circuit that caused the comparison of performances to favour the triode. It was as well that the Chief Engineer didn't notice the error! (To follow the problem through you would need some understanding of the dynamic resistance of tuned circuits and the associated mathematics.) No matter, it worked fine. The pentode version would probably have worked better!

It took a couple of years before a fully engineered airborne model of TACAN was completed. This might appear a long time. A large amount of the time was taken up preparing production drawings. The detailed and assembly drawings would run into several hundred, some of which would be very complex. Any error in any of the drawings could cause chaos in production.

Flight evaluation was carried out by the Ministry of Defence at The Royal Aircraft Establishment at Farnborough. It performed the required function, but did not meet all the required design specifications.

The US version was in production and entering service in the US Air Force. An engineer was dispatched to the US to establish why we were having problems and to seek remedies. His findings were not as expected. He was not particularly welcome at our parent company's development site – ITT Federal Labs in New Jersey. When he did eventually get access to the production test area he found that their equipment was no better than ours.

He would ask, for instance, 'How is it that this equipment is being shipped when the transmitter power is below specification?'

'Oh we've got a waiver on that,' would come the answer.

'How is it that the receiver sensitivity is low?'

'Yeah, we have a waiver on that.' And so it would go on.

We asked for a relaxation of the production test specifications, at least for the first production batch. No way!

We were pawns in a political game being played out between two government departments. Following the war our trade balance was in a precarious state. So much so that foreign currency for holidays abroad, for those that could afford them, was limited to £50. Thus the purpose of anglicising and producing the American design was to avoid a dollar drain. The Treasury would not sanction an American purchase. The Ministry of Defence wanted the original American equipment. Amendments to the design to accommodate British components could only be second best. By insisting that the equipment met the design specification as written into the contract, the MoD were able to force the Treasury's hand.

The production contract for the British airborne equipment was cancelled and the MoD got the American-produced units. The British ground and ship-borne equipment was allowed to go ahead. The environmental requirements of ground-based equipments are less stringent than those for aircraft equipment.

Another factor was the promising development of transistor technology. Transistorisation of the TACAN airborne equipment could solve the temperature and weight problems associated with the valve equipments. ITT Federal Labs were developing a transistorised version. Under NATO guidelines this could be a NATO approved unit. The transistorised unit would have made British TACAN obsolete as soon as it got into service.

Work was not necessarily exclusive to a single project. Other investigations could be going on concurrently. Some ended in disaster. We were tasked to investigate the viability of an airborne atomic clock, such as is now used for the international standard of time. Only one had been made. Birchenough and I visited the prototype at the Post Office research establishment at Dollis Hill. The caesium clock now replaces the astronomic method of defining the international standard of time, namely the second, and is based at Herstmanceau. It deviates less than a second per year, a stability of around one part in ten thousand million.

The purpose of developing an airborne standard was to provide an accurate aid to navigation. The reason the investigation was tasked to Birchenough was his practical knowledge of microwave technology and the similarity to the frequency multiplier chain that was required to the one developed for the TACAN equipment. I developed a stable oscillator source and associated constant temperature oven. In order to come into range of the spin frequency of the caesium nuclei a short-term stability of one part in one thousand million was required. This was obtained and we were ready to couple the oscillator source and associated frequency multiplier for comparison to the caesium tube frequency. A difference between the two frequencies would be fed back to stabilise the oscillator.

The caesium tube had to be operated in a strong magnetic field. Very heavy gauge wire was wound round horseshoe-shaped iron and clamped across the caesium tube. A lorry starter battery was to provide 100 amps for the magnet's coil. The cable from the battery was coupled to the free ends of the coil by a large brass ferrule. The circuit was to be closed with a heavy-duty lever switch. I closed the switch.

With all the high technology the basics had been forgotten. Every schoolboy knows that a changing electric current in a magnetic field results in motion. A very strong current in a very strong magnetic field created extreme motion in the wire carrying the large brass ferrule, which crashed into the caesium tube sending glass and the innards of the tube all over the bench. The tube was a one-off made in the research labs of the valve division.

The project was quietly closed, I think with some relief. It was not really practical at that time, and it was not a funded project. It was being done 'on the side.' A similar system was later developed elsewhere, after it became more realistic using integrated circuit technology.

My next project took me through to the end of my time at STC, and was to give me a great sense of achievement.

CHAPTER V

Blind Landing – Introducing Radar Altimeters

It was not, and still is not possible, to drive cars, or indeed any mode of ground transport, in zero visibility. It was fairly visionary therefore to believe that aircraft could be landed in zero visibility. Yet of such importance was this requirement that the Ministry of Defence set up a Blind Landing Experimental Unit (BLEU), first at Martlesham Heath in Suffolk, and later at an outstation of RAE at Bedford.

The need for the RAF to have blind landing capability was to enable the V bombers, which carried the nuclear deterrent, to be able to carry out a mission whatever the weather. They would have little fuel reserve to divert on their return.

Civil aircraft also needed to be sure of landing at their designated destination rather than divert with a full passenger load to an alternative airport that could be several hundred miles away. Apart from the benefit to the passengers it made economic sense.

A beacon approach system similar to that developed at TRE during the war was adapted to give the correct approach path.

STC with the parent company, ITT, was closely involved in developing aircraft approach systems, supplying both ground and airborne equipment for the internationally agreed instrument landing system (ILS). This was similar to the beacon approach system developed at TRE. It was to be the basis of the blind landing system.

To control the rate of descent over the undershoot and runway, very accurate knowledge of height above the terrain immediately below the aircraft was required. Barometric altitude related to height above sea level, and had a resolution of around 100 feet – not good enough. An altimeter capable of resolving height to one foot during the final flare out to touchdown was required.

BLEU looked into the possibility of using pulse radar. This could not resolve height down to zero since the ground return echo would run into the transmitted pulse. A continuous wave system with a continuously varying transmitter frequency could solve this problem. Such a system was flight evaluated at BLEU and found to be able to resolve the aircraft height to the point of touchdown.

An STC frequency modulated continuous wave radio altimeter was used for the early development work at BLEU. It used valve technology. A contract was placed on STC to develop an upgraded version using transistor technology where possible. For me this was a new door opening.

During my first five years at STC I had studied in the evening and attended various advanced technology colleges in order to obtain recognised academic qualifications. I became a chartered engineer, member of the Institution of Electronic and Radio Engineers (later absorbed into the IEE), and was elected to the Royal Aeronautical Society. I was the prize winner for my year in electronic measurements.

It was soon after this that Capelli called me to his office and asked whether I would like to take over the Radio Altimeter Development Group. I would. But I knew nothing of radio altimeters, having always worked on pulse systems. This was a frequency modulated continuous wave system. There were engineers who had worked on the original altimeter for several years. Furthermore, John Birchenough had been responsible for the design of the microwave transmitter and receiver. What would he make of my appointment?

He couldn't have been more helpful. That doesn't mean he was happy about the situation. After all I had, in effect, been his apprentice. A few months later he left to join another company. Another senior engineer in Capelli's group also felt he had been passed over and left as soon as he found another position. I was to come across both engineers later and was pleased that both were happy in their new posts.

I could not have been more fortunate in the engineers I inherited. John Purchase in particular had worked with John Birchenough on the microwave sources and receivers for the radio altimeter. He had inherited Birchenough's design integrity. I knew I need not question any design carried by John and could seek his opinion on any microwave requirement.

It was only a few weeks after I took over the radio altimeter group that Capelli asked me to take a look at the installation requirements for a new aircraft under development at the Blackburn Aircraft Company (later absorbed into BAC). This was the Blackburn Buccaneer, designed as a low-level ground following aircraft.

At that time the Soviet Union planned a fleet of 24 Sverdlov class heavily armed cruisers displacing an impressive 17,000 tons, with missile launchers and 32 anti-aircraft guns they could achieve a speed of 34 knots and had a fire control system superior to anything in the West. They actually put a dozen in service. To counter this, Naval Requirement NA.39 was drawn up for a strike aircraft capable of carrying nuclear and conventional stores, flying at 200 feet under enemy radar at 550 knots with a radius of more than 400 miles, carrying a weapon load of 4000 lbs and able to detect enemy shipping. The primary weapon would be a guided nuclear bomb called Big Cheese delivered by toss delivery and also a free fall device named Red Beard being developed for the Royal Air Force. This was all quite a tall order. Low-level attack was still a revolutionary idea and the RAF was sceptical.

This requirement was circulated to a number of British companies and Blackburn Aviation was awarded the contract. This was quite surprising in that the manufacturer was only producing the Beverley transport – hardly cutting edge.

The radio altimeter was to define ground clearance to enable it to follow the ground contours. The radio altimeter was to be the Mk7, for which I was now responsible. A problem arose in that there was no flat area under the aircraft to fit our horn antennas. I took the relevant drawing back to show John. He devised a contoured horn to fit. This was not a simple mechanical problem, as the aperture of the horn had to be electrically matched to the free space involving quite tricky measurement techniques.

There was a similar problem on the Vulcan aircraft. An aerial installation developed by the Blind Landing Experimental Unit indicated the direct distance between the transmitter and receiver antennas instead of the distance to the ground. It fell to me to tell the chief superintendent of BLEU, a PhD no less, that his installation was no good and would have to be replaced. Again, special contoured antennas developed by John Purchase solved the problem. My relations with the chief superintendent became strained and I was glad when he moved on.

Thereafter we established a good rapport with the establishment. This was just as well. The Radio Altimeter development was being funded by the Ministry of Defence and BLEU were the design authority. It had to work in their aircraft, which meant flight trials. I was welcome to participate in them. Measurements of signal characteristics encountered in flight were vital. The Avro Vulcan was a delta wing subsonic jet bomber that was operated from 1953 until 1984. The Vulcan was part of the RAF's V nuclear bomber force and was also used in a conventional bombing role during the Falklands conflict.

The civil requirement went on in parallel. British European Airways were to be the first to operate aircraft with automatic landing capability. This was to be fitted with an autopilot developed for 'autoland' by Smiths Industries. Smiths had a fully instrumented Varsity aircraft which was used for the initial trials until the designated aircraft, the BEA Trident, became available. The Trident was under development by Hawker Siddeley at Hatfield (previously de Havilland). The whole system had to be approved by the Civil Aviation Authority (CAA). Thus there was continuous liaison between all these organisations. Sometimes representatives from all the bodies would be on the flight trials. This helped when we had formal progress meetings. We appreciated that we were breaking new ground and all had problems to overcome. The civil aviation department, then part of the Ministry of Technology, had to provide the necessary ground equipment, such as ILS beams and lighting.

My main problem was 'double bounce'. The basis of a frequency modulated radio altimeter is very simple. The transmitted frequency is moved up and down at a constant rate. The difference between the transmitted and received signal is proportional to the distance from the ground below. Double bounce occurs when the primary ground return signal fades for various reasons coinciding with a strong signal reflected back from the underside of the aircraft back to the ground and back again to the receiver antenna as a double bounce signal, indicating to the autopilot that the aircraft is twice its true height. The autopilot would react by giving a nose-down command. The problem is fundamental to the propagation of radio waves in free space.

How to overcome the problem had eluded us for some time, yet it was vital. I developed and patented an 'electronic flywheel', analogous to a mechanical flywheel, that would not respond to transitional changes of input. This solved the problem. Simultaneously, our pure research department at the Standard Telephone Laboratories – called in to look into this seemingly fundamental problem – came up with the novel idea of looking at the inverse of the difference frequency between transmitted and received signal, and only taking note of the longer periods, which represented the shortest distance. This circuit was incorporated into the production units. Circuits were given unique identities, such as the multivibrators and flip-flops mentioned earlier. This new circuit was christened HATE for High Answers Totally Eliminated.

This was only one of the problems that had to be overcome to ensure the necessary integrity. The acceptable safety criterion for landing accidents was one in a million. It might be thought that none would have been the aim, but in real life it is recognised that with the best will in the world accidents do happen. The Civil Aviation Authority in the UK in conjunction with the international safety authorities set the acceptable level for autoland at one order higher than manual landings; that is on in ten million.

For the final part of the landing, information from the radio altimeter determines the rate of descent to touchdown. Not only is it necessary to eliminate all design problems, but to introduce high integrity production and test procedures as well. I introduced tests that simulated the 'free space' propagation delays that were encountered in operational conditions.

An air-conditioned clean area was set up for production. The completed radio altimeter unit was housed in a sealed canister pressurised with dry nitrogen to prevent contamination in the aircraft environment. The operatives, testers and inspectors had to wear gowns, hats and slippers similar to hospital operating theatre clothing. This was not to everybody's liking. Thus the radio altimeter production area was staffed by volunteers and new recruits. I christened it the 'United Nations' due to the broad spectrum of ethnicities present. I was called in if there were any problems (the buck stopped here) and had to don hat and gown likewise.

The first aircraft in the world to land fare-paying passenger automatically, without help or hindrance from human hand, was to be the British European Airways Trident aircraft. To obtain the necessary integrity the entire autoland system was triplicated, starting at the three engines. Each engine had its own electrical generator supplying independent power lines to the three autopilots, radio altimeters and other associated equipment. All three systems had to give the same output commands. If one differed it would be disconnected and the remaining two systems compared. If they differed the pilot would take over and make the decision to either continue the landing or overshoot and divert to another destination.

The final overall system approval for use in service was not without a bit of drama. I would not normally have been involved in the procedure, but I received a call early on the morning of the final approval flight. The trial's Trident aircraft had been fitted with pre-production radio altimeters. The Civil Aviation Authority inspector insisted on production units. The aircraft was standing by at Hatfield ready for take off. Units released into stores after final test should only be released from stores with the appropriate shipping documents. Shipping documents were raised by marketing. Marketing by their very nature would wish to be seen to be 'involved'. To go through that royal red route to raise the documents would take hours of explanation and fighting the internal bureaucratic machinery. That was not on.

I was an authorised signature for stores items. I signed a stores chit for three radio altimeters identified by the appropriate reference number. To the storeman one number was just like any other. He had an authorised signature and handed me the units. I took them to my car, put them in the boot and drove out of the gate with about £90,000 worth of equipment at today's prices. I was expected at Hatfield and drove straight to the hangar and installed the units under the watchful eye of the CAA inspector. I was invited to join the approval flight.

I had flown in the trials aircraft many times in order to take measurements and observe the altimeter performance. On this flight I was nothing but an interested observer. The approval was to take place at RAE Bedford using their long

runway with its associated kine-theodolite and other appropriate instrumentation. A visor was put down over the pilot's
and co-pilot's eyes as the aircraft was set for automatic landing.
After one or two abortive approaches during which one of
the three systems threw out, the Smith's engineer responsible for the autopilot made a couple of adjustments and several
'touch-and-goes' were successfully completed. The CAA man
expressed himself satisfied.

It was too late to return the production equipments to
stores, so they stayed with me for the night. The following
morning I booked the production units back into stores. It was
a fortnight later that the administrative system reacted. I had a
call to the accountant's office.

'What are you doing drawing valuable production units from
bonded stores?' he demanded. I explained the special circum-
stances and pointed out that they were returned the following
day. But accountants have one-track minds. I found that I was
no longer on the authorised signature list; a fine acknowledg-
ment for securing an order worth around £750,000. But then
I supposed that was what I was paid for.

In order to build up operational experience and reliabil-
ity data, all landings at airfields where ILS was available and
subject to the captain's decision were automatic landings. Thus
when the weather turned and visibility was reduced to zero or
near zero the 'blind' landing would be no different to the previous good weather landings. By the end of its operational life
the Trident had made close on one million automatic landings,
several in zero visibility conditions.

This 'majority vote' system seemed to me to have superb
integrity. Later systems used dual monitored systems. Only
one system would be connected at any time. The other system
would be on standby ready to switch on if the original system

monitor indicated a malfunction. To my mind this had less integrity. Whilst every precaution was taken to make the equipment 'fail safe', what would happen if there was a malfunction of the monitoring circuits?

Such a dual system was developed for the VC10 aircraft operated by BOAC. I was never happy with this system. Nor was I happy with the liaison with the development team led by the overseas division. In the event it had a comparatively short life. It was a very popular aircraft with the travelling public, flying on the Atlantic routes with a full passenger load when the American airlines were flying half empty. Passengers – American passengers – reckoned that the VC10 was the quietest and best aircraft in the air. Some American airlines were contemplating buying VC10s.

The story goes that when the newly elected Prime Minister Harold Wilson approached President Johnson for around £3,500,000 to enable him to carry out his social policy, the President pointed out that he could hardly bankroll the country that was endangering the profitability of the US.

Wilson appointed his own man, an accountant I believe, as head of the BOAC, which was government-owned at that time. He ordered Boeing aircraft to replace the VC10s on the basis that the Boeings had better fuel efficiency. This was true, but ignored the fact that a full aircraft used less fuel per passenger than a half-empty one. No other airline would now consider buying the VC10 if its own government had turned it down.

At about the same time as the demise of the VC10 I was visiting the General Dynamics Aircraft Company in Fort Worth. The British Government were contemplating purchasing the F111 aircraft and we were designing a special radio altimeter for the British purchase. The American

engineers were amazed that we had discontinued the development of our own swing-wing aircraft, the TSR2. They had
nothing that could compare. Despite a plea from an aircraft
museum the TSR2 prototypes were bulldozed and the drawings were destroyed! In the event the British government did
not go ahead with the F111 purchase. They did, however purchase the McDonnell Phantom fighter.

CHAPTER VI

The Phantom Project and the Radar Altimeter

The government contract on McDonnell for the Phantom fighter aircraft to replace the ageing English Electric Lightning was for the complete package; that is, McDonnell would be responsible for supplying the aircraft completely fitted out and meeting all the flight requirements. Thus the contract on STC for the radio altimeter was placed by McDonnell. This was to have important and interesting repercussions.

The specification for the radio altimeter was upgraded to incorporate the latest technology. The Klystron microwave transmitter was to be replaced by a solid-state source, and the original 'cat's whisker' receiver mixer diodes replaced with a state of the art receiver. The solid-state source was to be developed at the valve division at Paignton. This upgraded radio altimeter was designated the STR70.

The marketing department saw this new development as an opportunity to break into the world market, a market that historically they were dominant in – communication and navigation equipments. Their standing was threatened by the lead the Americans had established in semi-conductor (silicon chip) technology. There was a requirement for all RAF aircraft to be fitted with radio altimeters, a very prestigious market that would lead to worldwide sales. With this in mind trials were initiated with the Ministry of Defence.

Flight trials for the RAF 'autoland' requirements had always been carried out at RAE Bedford, with whom we had a good relationship. The STR70 trials were to be carried out at A & AEE Boscombe Down. I rigged up a lab prototype unit using an early version of the solid-state microwave source and had it fitted to the company Dakota aircraft. The unit employed the techniques that had proved successful for the autoland radio altimeter – in particular a fixed frequency excursion and rate of frequency modulation, albeit a fairly low rate to keep the bandwidth down. This was demonstrated to the MoD using their kine-theodolite range at Boscombe Down. The demonstration proved satisfactory. The MoD were to set up a full-scale trial of a finalised pre-production unit.

Such was the interest at the top level at the head office that a more senior engineer was appointed to take charge of this development. I was still heavily involved with the autoland projects, and a new installation for the Swedish Viggen aircraft. To my dismay the engineer appointed to look after the STR70 development decided to 'improve' it. Instead of a constant frequency deviation, it was progressively reduced between 500 and 5,000 feet. Whilst giving some improvement at low altitudes it became more vulnerable at high altitudes where the return signal was weakest. It also meant there was no longer a linear relationship between height and output. I can only imagine that he was trying to justify his appointment.

It was unfortunate that at this time C.E. Strong, a chief engineer of great integrity, had retired. This was 'the swinging

60s' and the era of the whiz kids, young new graduates with management qualifications but little experience. They didn't last long, but not before they had wreaked havoc. My protests were taken to be sour grapes.

The flight trials at Boscombe Down were in trouble. I was put back into the loop to try to sort out the poor performance but it was too late. This was compounded by a rival for the across-the-board-fit for precision radio altimeters. It was not that their equipment was superior – it could not work down to zero height, though in turn our unit was sometimes out of specification at maximum range. However, it was their marketing strategy that eclipsed ours. I could not understand the cool and unhelpful reception I received at Boscombe Down. The attitude was 'Forget it, your equipment is no good.' Unknown to me their mind was already made up. The services trials officer finished up as number two to the marketing director at the rival company.

During this time I continued with the technical liaison at McDonnell. They held progress meeting every six weeks at their location in St Louis. The novelty of visiting the States had long worn off. It was a bore. The meeting could be particularly long winded. I recall, for instance, discussions relating to the antenna (aerial) installation.

'We'll call in the stress man,' announced the chairman. He duly arrived. He spent about half an hour explaining his position in the organisation, complete with family-tree diagrams on the whiteboard, stressing how important he was. He then looked at the problem and said he would report back on the following day. An action was minuted to that effect; and so it went on. Eventually the McDonnell Phantom completed its trials at Edwards Air Force Base.

Whilst this was going on the Honeywell radar altimeter had been selected as the across-the-board fit for all RAF aircraft.

The Ministry of Defence requested McDonnell to retrofit the Honeywell radar altimeter in the Phantom. 'No way,' said McDonnell, 'the aircraft has met your specification in accordance with the contract, and that's that.'

The Phantom was in service in the RAF for the next twenty years. There were no problems with the radio altimeter.

CHAPTER VII

The Swedish Affair

In Sweden we had a technical and moral victory over our adversary at Boscombe Down. It turned out to be a hollow one.

The Swedes were developing their Viggen aircraft, a deterrent to any adversary who might consider aggression, in particular nuclear aggression. Russia had developed a nuclear bomb and was too close for comfort. The Viggen could be hidden from sight away from conventional airfields and was capable of taking off from a main highway. There was a requirement for accurate low-level terrain clearance information so it could fly below radar surveillance.

Three companies tendered for the contract: Bendix with a pulse system; Honeywell with a pulse system; and STC with a continuous wave frequency modulated system. Flight trials to check the feasibility of each system were to be held at the Swedish equivalent of RAE Farnborough. A date was set for the start and finish of the trials. There was no flexibility with regard to the date.

Time was short. I modified a civil version of the Mk7 radio altimeter. The new STR70 (see above), was still under development and in trouble. I kept to a fixed modulation depth, as for the autoland radio altimeters, to avoid the troubles of the STR70.

I arrived at the test centre on the appointed day and checked over the equipment before handing it over for installation in the trials aircraft. The first flight of the evaluation was made on the following day. The height data from the altimeter was evaluated against kine-theodolite readings. It worked fine within the operational range, but continued to give readings when above maximum range. This was unacceptable. The problem was fundamental to the antenna sighting, which could not be changed. Somehow I would have to deal with it in the signal processing circuits.

I worked continuously the next day and into the evening without breaking for meals. I was very conscious that I alone could secure or lose the contract. The Viggen project manager took a keen and jovial interest in what I was doing and offered all the help and support he could. I explained that I was adding a second error detector to the HATE signal processing circuit to look for variations in the periods of the counter frequencies. He was optimistic, showing greater confidence in the outcome than I felt. If it worked I knew we were home and dry. I knew the Bendix equipment, by its very nature, could not work down to zero, and the Honeywell equipment didn't show up.

It worked. Some weeks after my return STC were offered the contract. A high-powered delegation of senior management flew out in the company aircraft to negotiate the contract. The Swedish project manager looked round at the assembled gathering and asked, 'vere iss Mr Goult?' A telegram was sent requesting me to catch the next plane. This sort of thing does not go down too well at home.

Meanwhile our old protagonists were champing at the bit after missing the trials and were hammering at the door trying to get another chance. They went a step too far. The story goes, so I heard through the grapevine, that the Swedish government were offered the results of the Boscombe Downs trials of the STR70. This was confidential information and in contravention of the Official Secrets Act. The ploy back-fired. The Swedes reported back to the British government. I am not party to the consequences, but some time later I had reason to bump into the ex-trials officer from Boscombe Down. He appeared to have become an alcoholic and looked a nervous wreck.

The Swedish affair had a sting in its tail. ITT, our parent company, decided to close the Radio Division of STC. We would not be able to fulfil the contract. STC had to pay a penalty for defaulting on the contract. It opened the door to Honeywell.

It was another door closing for me, but before it did there was an interesting interlude. I recall it below as I remember it. It was a long time ago!

CHAPTER VIII

Russian Interlude

I looked at the pale young lady – or girl, I could not make up my mind which – who had just been introduced. It was a formal introduction, so I shook hands with her and sensed her nervousness.

'I have never spoken English to an Englishman before,' she faltered in an almost inaudible voice.

'You speak it very well, Galya' I said.

'I am glad you find it so,' she responded with obvious relief.

I was part of an international conglomerate taking part in a conference in Moscow, my first visit to Russia. There were about thirty in all from the headquarters in New York and associated European houses. I had written a paper to be read at the conference. It had already been translated into Russian and distributed to all the attendees.

'I am to translate all the questions from Russian to English,' continued Galya. 'Then you can write the answers in English and I will return your answers to the questioner in Russian.'

'It sounds a long winded process.'

'Long winded?'

'Er, drawn out, tedious. I can't think of the right word.'

'I know what you mean. We shall see. I have some questions all ready for you.'

'Already! I didn't think there would be any questions until my paper had been read, and that's another two days yet.'

'The engineers have had the papers for a week now and I received the first questions this morning. I can translate here when this reception is over.'

We were in a large anteroom leading off the conference hall. The visiting delegates were meeting their hosts. There was a translator attached to each of the visiting delegates, all in earnest conversation. The majority of the translators seemed to be young students or graduates, but the more senior delegates such as the technical director and vice president had interpreters attached to them. Some delegates were gathered round an urn-like object.

'What is that?' I asked.

'A samovar. Would you like some tea?' She led the way towards it. She had regained her confidence.

'I have never seen a samovar before. Do you keep it going all the time?'

'Of course. You will be able to have hot tea at any time during the conference.'

'Great, I shall need constant sustenance coping with all those questions you have for me.' She gave ghost of a smile. She was very attractive when she smiled.

Others joined us at the samovar and soon a small informal group was formed, a colleague from STC with his young translator, and Pierre, from the LMT, the French subsidiary of ITT. I apologised for my poor French.

'Don't worry,' replied Pierre in perfect English. 'You British are all the same. You can never speak any language but your own.'

'I only speak a little English,' said Pierre's translator slowly with great deliberation.

'Never mind,' said my colleague, 'perhaps I can improve your vocabulary.'

'Not a chance,' Pierre responded. 'I shall be helping her to perfect her French. Not that you don't speak it well already,' he added hastily in French. I smiled; my colleague had met his match in the suave and elegant Frenchman.

'It is time we looked at these questions,' said Galya.

I found it pleasant working with Galya, so serious, so precise. She handed me each question neatly written on separate sheets, each with its own identity number.

'You understand the question?' she asked.

'Oh yes, I understand the questions all right. It's supplying the answers that is the problem.'

'Surely you are a specialist in the subject. You would not be here otherwise.'

'You know what they say about specialists. They know more and more about less and less.'

'Pah! We say that here too. It is not true.'

'I suppose you specialised in languages.'

'No, no, I am an engineer. I learnt English at school.'

'Remarkable! How did you become so fluent?'

'Reading and listening to the BBC. Books that are banned in Russia can still be obtained from the English library. I have even read *From Russia With Love* by your Ian Fleming. There are many marvellous English authors. Who could match Charles Dickens, for instance?'

'Who indeed?' I thought it was time to change the subject. 'Do you enjoy engineering?' I asked.

'No. I was steered into the sciences at school because I was good at mathematics. I would rather have studied the arts.'

'So why don't you take up a career more orientated to the arts?'

'Because I am an engineer.'

In communist Russia there was no answer to that. 'I had better get on with these questions,' I said.

After my paper had been read the pressure was off. So long as any remaining questions were answered it seemed pointless hanging about the conference centre, so the little group that had formed went on sightseeing tours of Moscow with the young translators as guides.

We visited the Kremlin and were able to go inside. Galya pointed out that the walls simply enclosed the old city. The main government buildings were within the walls and were modern and well guarded, but many of the original buildings remained.

The Armoury, which was in fact a museum, contained all the old Tsarist treasures – diamond studded saddles, swords presented to past Tsars by Sultans and other world rulers, a massive coach on runners in place of wheels for fast winter travel. We were supposed to admire the priceless Fabergé Eggs – decorative eggs with embedded diamonds. Why anyone should be pleased with such a useless gift I could not imagine.

At the weekend there was an official visit to Leningrad. I was prepared to be bored, but Leningrad was a beautiful city owing much to the inspiration of Peter the Great. A statue stood of him astride a horse his drawn sword pointing towards Sweden as if daring the Swedes to return. Both the Winter Palace and the Hermitage Museum were built at his behest. The art collection, containing examples of all the masters was so great that in one afternoon it was only possible to view the best of the best.

I was getting a new perspective of the world, looking west from the east. They had been attacked from the west for centuries. Napoleon had entered Moscow. Hitler had reached its outskirts. The siege of Leningrad had reduced its population by over 100,000, many freezing and starving to death. If I were Russian I would find the massive show of strength displayed in the Red Square each year greatly reassuring.

Back in Moscow, after an absence of only two days it was like a reunion of old friends as we gathered once again in the conference hall. A full week of activities had been mapped out by the women.

The two weeks in Russia seemed like another world. Whilst with STC I had visited many countries, but it was the Russian visit that stays in my mind. So much so that I used it as the basis for a novel.

PART III

Avionic Systems
(Heathrow) Ltd

CHAPTER IX

Getting Started

'Three thousand redundancies, that's the whole of the radio division,' I said.

'Not much hope for us at our age,' said Bob, pushing away his unfinished soup. We were in the senior staff-dining room at STC. The food was no better than in the canteen, but it reflected the senior level of those privileged to eat there. It had had an air of distinction when presided over by C.E. Strong, a revered and respected chief engineer, to whom excellence in engineering, as in all else, was sacrosanct. Those days were gone. These were the days of the professional managers who had been put in by the parent company. The place was closing down.

I agreed with Bob. We had been with the firm for twenty years and believed we would finish our working life there.

'I know what we'll do,' I said.

'What will we do?'

'The same as we're doing now.'

'How's that?' came his unenthusiastic response.

'You are manager of the service depot at Heathrow, aren't you?'

'I don't deny it.'

'And isn't it one of the few flourishing and profitable areas left?'

'A lot of good that does me,' he grumbled. 'They want to flog it off to that dreadful outfit near Heathrow and damned if I am going to work for them.'

'Exactly,' I said. 'We'll buy the place ourselves.'

The silence was deafening.

'And,' I added for good measure, 'my own project isn't doing too badly. They could throw in the radio altimeter design rights as well. The Civil Aviation Authority would welcome it. They are worried about continuing support for the autoland project. The MoD are anxious about Post Design Services on the whole range of airborne equipment. We could cover this. Or am I indulging in fantasy?'

'No. I was reading in the paper only yesterday of some redundant managers who had bought part of their old company and continued trading. The journalist was impressed, thought it might catch on. There is,' continued Bob thoughtfully, 'just a little matter of money.'

'Yes, I have noticed that it, or rather the lack of it, tends to get in the way whenever you want to do anything.'

'It will need a bit of thinking about,' said Bob.

It took less than an hour to get started. Bob rang his insurance broker, and I rang head office.

'How did you get on?' I asked over the telephone. Bob had gone back to his office at the Heathrow depot. He was manager of Customer Services, responsible for maintenance of all the divisions' products, mainly aircraft navigational and radio communication equipment. I was located at the radio division in New Southgate, in charge of several engineering development groups.

'I had an encouraging chat with the broker,' Bob said. 'He said it should be looked at as a property deal. He will get in touch with the appropriate building societies. He will expect introductory commission for arranging the finance and all the insurance will go through his office. Oh, and he did say we needed to get full details of the buildings lease, conditions of trading and sales forecasts to establish credibility. How did you get on?'

'I discovered that the marketing director, Rex Clarke [not his real name], has responsibility for the disposal of all aspects of the division. I managed to get through to his secretary. She promised to pass on our interest, but I will write a formal letter and ask for all the details.'

My secretary, I considered was the best in the division. She smiled after I had finished dictating my letter.

'Things should get interesting,' she said.

A week passed. 'Isn't it time we heard from head office?' asked Bob, looking moodily out of my office window. My office was his last port of call before returning to his own location at Heathrow. He made a weekly visit to the New Southgate division to discuss servicing problems with the various design engineers and to try to extract more money from the financial controllers to replace his aging equipment, not that there was much hope of that.

'I'll give Clarke's secretary a buzz,' I replied lifting the receiver and dialling a number that was etched in my mind, for I had dialled it many times before.

'Can I speak to Mr Clarke?' I inquired of the usual detached voice.

'I'm sorry, Mr Clarke is busy,' came the usual reply.

'Busy doing what?' I persisted.

'Planning the disposal of your division,' came the haughty reply.

'That is exactly what I wish to speak to him about,' I said. 'I wrote him a letter last week formally expressing our interest and requesting information regarding the sale of the service depot. We've had no response.'

'I'll let him know you rang,' she said with an air of finality and replaced the receiver.

'They're not taking us seriously,' I said, turning back to Bob. 'If McKenzie was still managing director I could have talked to him, but I've never met this new guy, although he must know of me. He has to give his OK when I want to borrow the company aircraft for flight trials; I book through his secretary. I'll give him a ring.' Needless to say I got his secretary, but she was charming.

'I'll tell you what I will do', she said. 'Rex Clarke has just looked into the office. I'll call him on the way back and get him to talk to you.'

'I had hoped to talk to the managing director himself,' I said, disappointed.

'He's delegated the whole thing to Rex. It would be better to talk to him direct.'

Rex Clarke came on the phone about an hour later.

'I understand you are interested in making a bid for the maintenance depot and certain design rights.' I acknowledged this. 'Very well,' he said, 'I will send someone to see you.'

And so it was that we met 007, a jolly little fellow who had once been responsible for security vetting, although we were not supposed to know that. I think his name also happened to be Bond. He had been in the sales office previously, so he was slightly known to us. This made our discussions fairly relaxed and informal, although we recognised that we were dealing with a messenger who could only report back, not negotiate. However Mr Bond took his duties very seriously and following his backslapping, hail-fellow-well-met introductions he advised us that he was the company's representative for the sale of the many and various parts of the radio division and all matters relating thereto should in future be referred to him.

He pointed out that we were only one of many who were making an offer for the valuable parts of the radio division, adding that we could expect no special consideration as company employees. We commented that we didn't seem to have received any consideration.

'No, no, no,' he exclaimed. Provided that we could demonstrate financial credibility we would be treated the same the same as everybody else. Could we tell him how we were going to finance the purchase? We told him we were getting financial advice.

Next Mr Bond produced a glossy brochure of the items of interest. Bob's service depot was presented as a perfectly situated site of over 5,000 square feet of office and workshop space including clean areas for the repair and test of special instruments, fully fitted to enable repair and calibration of aircraft navigation and communication equipment. There followed an accurate account of growth and profitability over the past decade and glowing prospects for the next decade.

'Fine,' said Bob, 'I didn't know I was so good. Do you realise I was completely ignored out here? Only one senior manager visited the sight and no director came near the place. Now I have got to fight to buy my own success; but a major factor has been overlooked. Profitability and growth have reached a peak. The order book is at its lowest for years. So where is the future growth you predict?'

'That is up to the new management,' responded Mr Bond, quite unabashed.

The story was the same for the radio altimeter design rights. A world market of up to two million pounds was predicted over the next ten years. This was more than had been achieved over the last twenty years. Our Mr Bond's response was the same.

The first thing we needed was details of the lease. Obtaining of a copy of this was a saga in itself.

'Why did we want a copy?'

'To enable a proper evaluation.'

'The valuer will have to apply to our legal department.'

'So why can't we apply to the legal department?'

'You can.' And so it went on.

When we eventually obtained a copy it didn't do us much good. It now became apparent that we would have to revise our ideas on the method of financing the purchase. Our broker simply dealt with property and insurance. He was just hoping he could get some business out of it. Our credibility with the company was at stake.

CHAPTER X

High Finance

If you have no interest in how companies get started, please skip this chapter. I hope it may reveal some things; and this moment was certainly an important one in my story, so I beg your indulgence in including it.

Having flirted with a potential American backer without coming to an agreement, success, when it came, came from an unexpected quarter, and in an unexpected way. Accountants, in my opinion, can only run a business down, never build it up. They only know how to prune. It was A.V. Roe, De Havilland, Sopwith, Handley-Page, Henry Royce and other well-known names that founded the great British aerospace industry – all engineers. Whilst visiting the McDonnell Aircraft Corporation I was told it was the only aircraft company operating in the black at that time. The directors were all engineers. The founder still spoke with a Scottish accent. What a pity he had to go to the US to succeed.

Then there was Marconi, not the inventor of radio, as is generally supposed. Clerk Maxwell expounded the theory for the propagation of electro-magnetic waves at Cambridge and later

Heinrich Hertz demonstrated it in the laboratory. Marconi, inspired by the idea of worldwide wireless communication, experimented further to achieve this end. Furthermore he came to Britain to do it. Today he would go somewhere else. From Marconi's initial enthusiasm blossomed the vast radio and electronics industry of today, employing hundreds of millions of people worldwide.

When Ted Heath was Prime Minister he made money cheap by encouraging low interest rates, with the object of encouraging industrial and entrepreneurial investment. So what did the financial whiz kids do? They ploughed the cheap money into property at the expense of industrial development.

I would not equate myself to any of the great entrepreneurial engineers, nor was our proposed venture of earth shattering significance, but attitudes are reflected down the scale, and accountants in general, like the city operators, are not, to my mind, associated with entrepreneurial activity. Yet as it turned out it was two accountants who eventually got us going in business. They were unlike any accountants I have come across before or since. Furthermore it was a finance broker – a species we had written off – who gave us an introduction to them. We weren't sure how two accountants could help us, but by that stage we were clutching at straws. It was arranged that we would meet Simon and Peter at a pub in Twickenham.

The pub was not far from Twickenham rugby ground and on entering we saw a couple of well built rugger types at the bar behind their recently filled pints. Harrow and Oxford I would have guessed from the sound of their voices.

'You're probably looking for us' boomed one of them. Introductions were made.

'Can we get you a drink?' asked Simon, the elder of the two.

'Thanks' we said and Peter ordered four pints.

'Now tell us about yourselves,' said Simon as we settled ourselves at a table in a quiet corner. We explained our background and recent disappointments.

'OK,' said Simon. 'There's no point in taking on such a business unless you have a major shareholding. Don't do it for anything less.' That's all very well, I thought, but where do we get the cash. Simon was ahead of me. 'Where do you think we should go, Peter?' he said.

'What about Barclays of Kingsway?' suggested Peter.

'Good thinking.' Simon turned to us, 'I've done business with the manager before. I'll ring him tomorrow.' And so we parted, full of beer and renewed hope.

A few days later, by some apparent miracle we found ourselves in the manager's office in what appeared to be the most prestigious bank in Kingsway. An easy and relaxed rapport was soon established with the manager, who had obviously done business with Simon on many previous occasions.

'I can certainly support your working overdraft,' he said, when he had heard Simon give an outline of our proposal. The capital investment will have to come from elsewhere. You will need a merchant bank for that.' We told him that we had tried merchant banks, but they were not interested in small businesses.

'Try John Tomlinson [a pseudonym] of Ansbacher. He recently set up a large warehousing facility at Heathrow. Phone him. Mention my name and say I recommended you. You should send these cash flow charts and profit and loss predictions under your letterhead,' he added to Simon.

We didn't have to wait long. After about a week and some form filling Mr Tomlinson's secretary rang to say he had business at Heathrow and would take the opportunity to call and see us. He wasted no time wanting to be shown round.

Although he had no technical background he showed a ready appreciation of what was going on, asking question after question in quick succession, often not waiting for a complete reply before asking the next. Marching straight into goods inwards he asked without ceremony of the only occupant, 'What's that you are doing?'

'I'm batching components,' replied the surprised chief inspector. 'They will be given a batch number up to the time when they are assembled into equipment and the number will be recorded on the unit's job sheet or card so any component can be traced back to its source.'

'That's right,' responded Bob. 'This is our chief inspector.'

'How many inspectors do you have?'

'Three. The other two are in the workshop checking the work of the technicians and testers. Work cannot be released for delivery without the inspector's stamp and initials. The stamps are only issued to inspectors approved by the Ministry of Defence or the Civil Aviation Authority and are not transferable.'

'Does all your work need MoD or CAA release?'

'Yes.'

We adjourned to the office for coffee while our potential source of finance looked at the orders on hand. 'I see that British European Airways are one of your biggest customers,' he said. Is there anyone there I could talk to?'

'Paddy O'Hanlon of radio communication and navigation would be your best person,' I replied. 'I worked closely with his department when on the autoland project. I'll give him a ring.'

On learning who John Tomlinson was O'Hanlon invited him over – BEA Engineering was situated across the road from us. They spent 45 minutes together. John Tomlinson came back well pleased. 'Now can you put me in touch with the CAA surveyor?' he asked. Again, I had worked closely with the CAA on the autoland project and gave John Tomlinson the surveyor's number. He gave an assurance that if we took over he was confident we could obtain CAA approval. A similar response was obtained from the MoD.

'OK', said John, 'I'll back you.'

CHAPTER XI

We're in Business

Avionic Systems (Heathrow) Ltd – hereafter shortened to ASH – was engineer-led. We now settled down to a sustained period of growth. Most of the original staff stayed with us. The original operation at Heathrow was maintenance of STC equipment. We now added research and development to these activities. Development engineers from our old company were taken on to assist me with this programme.

We obtained two major contracts as a result of acquiring the radio altimeter design rights, one from the MoD and one from BEA. This led me into my first foray into microwaves. At STC I had microwave engineers in my development groups. Now I was on my own. The problem was to replace the reflex klystron tube used in the Mk7 radio altimeter transmitter, which had a limited life, with a solid-state device. I had been responsible for the design of this equipment at STC. Incidentally, during the war, the Axis powers relied mostly on low-powered klystron technology for their radar microwave generation, while the Allies used the far more powerful but frequency-drifting technology of the cavity magnetron for microwave generation. Klystron tube technologies for very high-power applications such as synchotrons and radar systems came later.

The solution was to fit a special diode (a Gunn diode) operating in its negative resistance region in a resonant waveguide previously excited by the klystron. This nearly ended in disaster when I dropped the sample Gunn diode generously sent to me by the manufacturer. It was little bigger than a grain of wheat, was the only sample I had to work with, and seemed to have completely disappeared. I was already working late into the evening in order to get the development completed by a tight deadline. It took me nearly an hour on my hands and knees before I located it. Late home again! Eventually the new Gunn diode microwave transmitter, together with an improved receiver pre-amplifier was completed, resulting in a substantial Ministry contract to upgrade their entire stock of Mk7s.

BEA were also concerned about the replacement cost of the klystron transmitter tube, the only lifed item in their autoland radio altimeters. We had been talking to BEA during our protracted negotiations to raise cash. Eventually they could wait no longer. They gave a study contract to a larger company. We thought we had missed the boat. As soon as we were up and running I wrote a paper defining our proposal. BEA showed remarkable faith in us and gave us a contract to develop and provide six models for flight trials. Depending on the result of the trials a production contract would follow. Furthermore, they gave us an advance for the development work and the promise of progress payments for the production contract on the successful completion of the trials. This was the largest contract we ever handled. Whilst it was a tremendous boost to our cash flow, surprising and delighting our bankers, it was

1 Robert (later Sir Robert) Watson-Watt and Arnold Wilkins, who, following an investigation of the 'Death Ray' in 1935, conducted an experiment illustrating the principles of radio direction finding of aircraft – the precursor of radar.

2 Bawdsey Manor, the site of the Air Ministry Research Establishment (later known as TRE) from 1935–1939. *(Courtesy Lance Cooper)*

3 Bawdsey Manor viewed from the Felixstowe side of the River Deben, 1938. Note the height of the aerial masts.

4 Charley Brinkley, who ferried staff across the river. It took only seven minutes, while it took 45 minutes by road via Woodbridge.

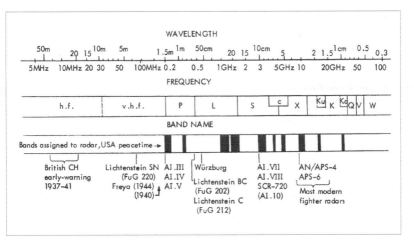

IN THE YEAR 1936 AT BAWDSEY MANOR
ROBERT WATSON-WATT
AND HIS TEAM OF SCIENTISTS DEVELOPED
THE FIRST AIR DEFENCE RADAR WARNING STATION.
THE RESULTS ACHIEVED BY THESE PIONEERS PLAYED
A VITAL PART IN THE SUCCESSFUL OUTCOME OF
THE BATTLE OF BRITAIN IN 1940.

5 The memorial plaque at Bawdsey Manor.

6 A radar's wavelength is the reciprocal of the operating frequency. 1MHz is one million cycles per second; 1GHz is 1,000 million. The portion of the electromagnetic spectrum used for radio and radar is subdivided into 'bands', each identified by a letter. Most modern AI radars operate in the X or K band.

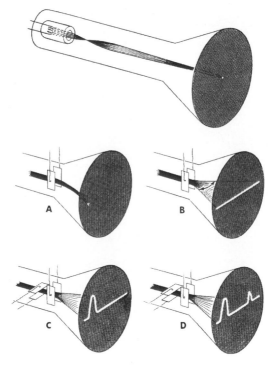

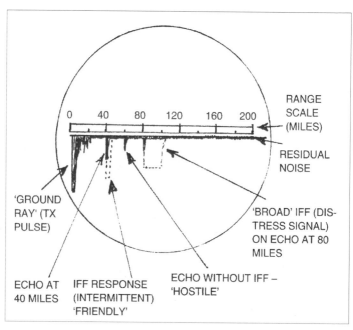

RANGE SCALE (MILES)

0   40   80   120   160   200

RESIDUAL NOISE

'GROUND RAY' (TX PULSE)

ECHO AT 40 MILES

IFF RESPONSE (INTERMITTENT) 'FRIENDLY'

ECHO WITHOUT IFF – 'HOSTILE'

'BROAD' IFF (DISTRESS SIGNAL) ON ECHO AT 80 MILES

9 The original airborne radar team, 1936–1943.

Opposite from top

7 The cathode ray tube (CRT) display. A hot filament sends out a stream of electrons (cathode rays) which are focussed by a magnetic field and made to impinge on a glass screen which is coated with a phosphor so that, when bombarded by the electrons, a glowing spot can be seen.

To make a simple radar the cathode ray tube is provided with two deflector plates (A) which, when an electrical voltage is applied, bend the beam. If a high-frequency signal is applied, the beam is made to oscillate along a 'timebase', causing a bright line (B). Adding two more deflector plates means the beam can be deflected in the vertical plane. If these latter plates are connected to a receiver aerial then a large 'blip' will appear as each pulse of radar energy is send out (C). If there is another aircraft in the sky ahead, a small portion of the pulse will be reflected back to the receiver aerial, causing a second blip at a point on the timebase corresponding to a known target range (D).

8 Cathode ray tube radar display, showing echoes and IFF responses.

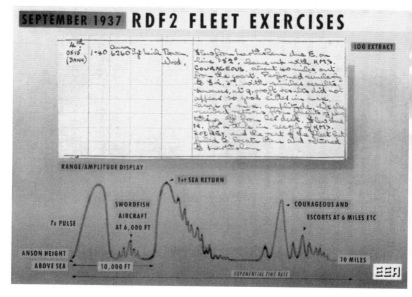

10 RDF2 Fleet exercises.

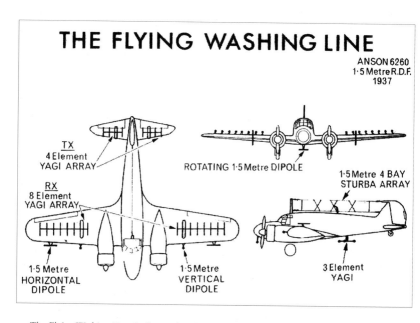

11 The Flying Washing Line. In September 1936 a suitably equipped Avro Anson detected aircraft carrier *Courageous* during a RN exercise at a range of 10 km – and then the interceptors she launched, the first aircraft detected on an airborne radar screen.

12 TRE at Worth Matravers, 1940. There were three other sites at requisitioned schools in Swanage and Langton Matravers.

13 Malvern College.

16 Two junior scientific officers with two lab assistants on the Malvern Hills. The scientific staff were generally youthful. Ted Ley, standing, invited me to take my first flight to observe the operation of GH.

Opposite from top
14 Some of the senior men of TRE. Top row: R.A. Smith and R.C. Cockburn. Seated: J.A. Ratcliffe, C. Holt Smith, R.A. Rowe and W.B. Lewis.

15 The office staff, hard worked but anything but sombre!

CONTROL
UNIT

RECEIVER

RANGE
CALIBRATOR

PPI
AZIMUTH
SCALE

INDICATOR
UNIT
HEIGHT/
RANGE
TUBE

SWITCHES

POWER
SUPPLIES

AUTO CUT-OUTS

INDICATING
NEONS

POWER
SUPPLIES

17 CGI receiver and display console.

Opposite from top

18 A typical 3cm H$_2$S picture on a navigator's screen.

19 Shipping off the Normandy coast on 6 June 1944 (D-Day) made what an H$_2$S operator would consider a clear picture. The aircraft is almost exactly over the coastline.

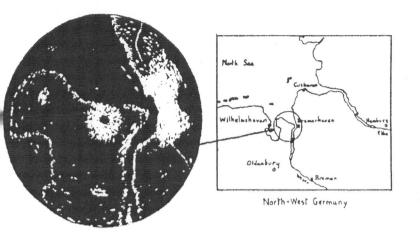

North-West Germany

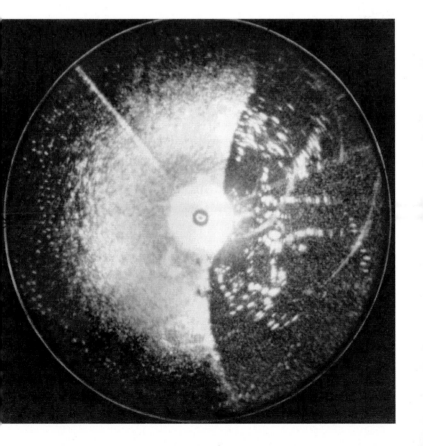

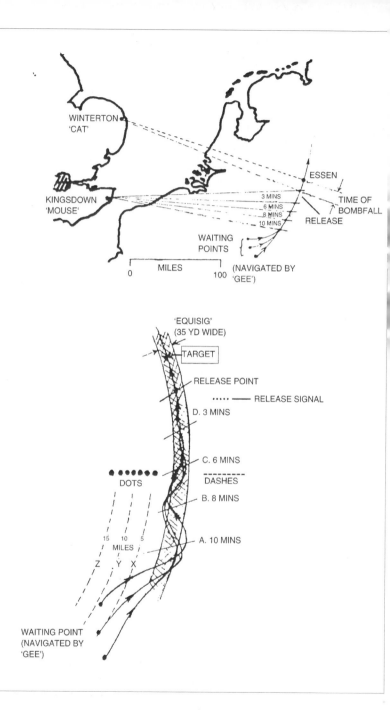

WINTERTON
'CAT'

ESSEN

3 MINS
6 MINS
8 MINS
10 MINS

TIME OF
BOMBFALL

KINGSDOWN
'MOUSE'

RELEASE

WAITING
POINTS

0   MILES   100

(NAVIGATED BY
'GEE')

'EQUISIG'
(35 YD WIDE)

TARGET

RELEASE POINT

••••• ——— RELEASE SIGNAL

D. 3 MINS

C. 6 MINS

•• •••••
DOTS

— — — —
DASHES

B. 8 MINS

15   10   5

MILES

Z   Y   X

A. 10 MINS

WAITING POINT
(NAVIGATED BY
'GEE')

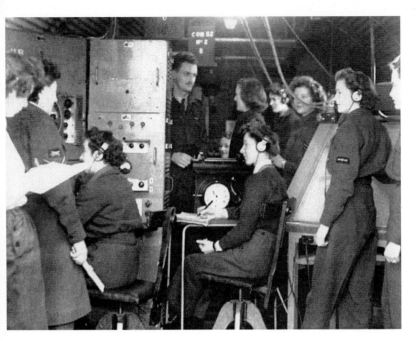

21 WAAFs setting up the target and time of bombfall in a UK Oboe station in 1943.

Opposite 20 Oboe: Two ground-based transmitters and receivers spaced miles apart would simultaneously transmit a short duration pulse. A pilot would be able to define his position according to a series of dots and dashes depending on his position in relation to the 'cat'.

22 A typical east coast CH station with four 360ft transmitter towers on the left and the 240ft receiver towers in the distance.

Opposite 23 Various configurations of 10 cm radars in RAF service.

CIRCULAR DISH ON
GANTRY OVER
NISSEN HUT

TYPE 52

HORIZONTAL
CHEESE ON
GANTRY OVER HUT

TYPE 53

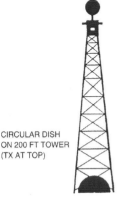
CIRCULAR DISH
ON 200 FT TOWER
(TX AT TOP)

TYPE 54 (a)

CIRCULAR DISH ON
TOWER. TX IN NISSEN
HUT AT BASE

TYPE 54 (b)

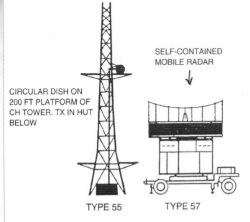
CIRCULAR DISH ON
200 FT PLATFORM OF
CH TOWER. TX IN HUT
BELOW

TYPE 55

SELF-CONTAINED
MOBILE RADAR
↓

TYPE 57

CIRCULAR DISH ON
185 FT WOODEN
TOWER. TX IN NISSEN
HUT BELOW

TYPE 56

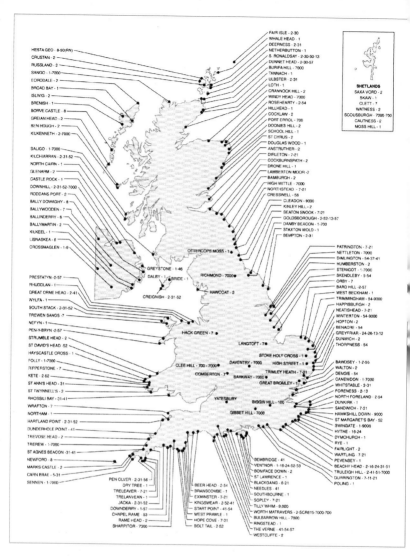

24 Locations of RAF radar and navaid stations during the Second World War.

also a hard fight – clearing design problem after design problem. We had to succeed. With each new problem the pressure built up. The deadline for the flight evaluations was absolute. By the laws of cussedness there was always a last-minute hitch as the deadline approached, but somehow we made it. We were awarded the production contract.

There was a useful follow on from the BEA contract. Following a series of accidents involving aircraft ploughing into mountains or into the ground in poor visibility, an international mandatory requirement was introduced requiring all passenger-carrying aircraft to be fitted with a Ground Proximity Warning System (GPWS). The BEA Trident was fitted with an automatic landing facility, as mentioned above. Three low range radio altimeters were associated with this system. We supplied an additional output compatible with the GPWS requirements. Self-monitoring was achieved by an internal delay line to simulate the delay of a ground return signal using a small portion of the transmitted signal. The delay time had to be within defined limits. The quality of the ground return signal was also monitored. The modifications were carried out in a clean area similar to the production area used for the original assembly at STC. The modified units received CAA approval and were fitted to the BEA Trident fleet.

We also had a very useful post design contract with the MoD. This required us to be responsible for any problems arising in service for the full range of ex-STC aircraft radio and navigational aids. This work was supported by a very co-operative work force.

CHAPTER XII

An Encounter with MI6

In an effort to obtain new business we made a couple of sorties behind the iron curtain. STC, our previous company, had sold equipment to Czechoslovakia. Perhaps we could resurrect some of it. We set off for Prague armed with some introductions from the London Chamber of Commerce and Industry. We were well and politely received by our contacts, all of whom expressed an interest in our proposals to carry on supplying equipment and services where STC had previously done business, although the approval of the appropriate government departments would be needed.

The city itself was depressing to see; the restoration of the beautiful bridges and buildings which had begun prior to the unsuccessful Dubjeck uprising in the 1950s had been stopped by the Russians in reprisal. Bob had visited the city during his earlier days at STC and attempted to look up some of his contacts. All but one had disappeared. He managed to contact this one at the offices of the Czech airline. Our meeting was held in the company of a young Czech engineer. After some discussion relating to test equipment the young man left the room.

'How goes it?' asked Bob.

'If it got better it would still be unbearable,' came the whispered reply.

At a reception at the British Embassy for us and other delegates it was noticeable how smart and prosperous looking were the majority of the Czech delegation contrasting sharply with the drabness that had settled on the city as a whole. Talking to one such official he told us that Czechoslovakia was a bureaucratically run country and that endless meetings would not necessarily result in any action. Money was in short supply, except for purchases from Russia. How right he was. Not one of our enthusiastically received proposals got past the bureaucrats.

Although we got no business from the eastern bloc it did result in an interesting episode. It was hoped that the seeds sown in Czechoslovakia might turn into firm business. With this in mind we took the opportunity to visit any function arranged by the London Chamber of Commerce that related to Czechoslovakia. It was at one such function that a first secretary, or some such, approached us and asked if we would like his help in promoting our activities further. 'We would welcome it,' we said.

'OK I will come and pay you a visit,' he said. We arranged a date to meet.

Serc was a young man, probably in his early thirties. He told us that he, his wife and baby rented a flat in St John's Wood. His English was very good with only a slight accent. He was anxious to help us capitalise on our Czechoslovakian visit. We went over our full product range with him. He returned after a couple of weeks, saying he had progressed the test units we had been promoting and would maintain contact with potential customers.

'There was some interest in your radar altimeters for measuring ground clearance,' he said. Bob said that was my area and passed him over to me. I showed him the civil equipments used in conjunction with automatic landing and ground proximity warning systems.

'Great,' he said. 'I am sure there will be some interest here. Have you the specifications they are made to?' I gave him a copy of the Declaration of Design and Performance, a document that is submitted to the CAA for design approval.

'Oh, by the way,' he said as he was leaving, 'I see you are a member of the IERE. I need to look up some references for a colleague. I tried the library of the Institution of Electrical Engineers, but not being a member they would not give me access. Do you think I could look them up in your institution's library?'

I phoned the librarian and she said it would be all right so long as I came with him and introduced him. I took Serc along one lunchtime and left him there. I think he ordered a few technical papers that had been published in the institution's journal.

About a month passed. Serc and Czechoslovakia had gone right out of my head when Serc phoned.

'I am a bit pressed for time,' he said. 'Could I possibly come and see you one evening? You don't live all that far from me and I have something for you.' I guessed he had at last turned up some business for us. He arranged come over at eight the next evening, and arrived carrying a heavy box.

'I remember you said how much you admired the Bohemian glassware you saw in Prague. 'So,' he said, unpacking a beautiful heavy blue tinted vase, 'I thought you might like this.' I was in a quandary. I would like to have refused, but this would have appeared rude. I took the easy way out and accepted it with all the appropriate words of thanks, but saying he should not have gone to all that trouble.

'No trouble at all,' he responded. 'It was a pleasure' and after a pause, 'I would like to say how impressed my colleagues in Czechoslovakia were with the performance figures of the radar altimeter. They wondered how you got such good resolution down to zero. Could you manage a short paper on how you achieve this?'

'Well,' I hesitated, pondering that you never get owt for nowt as my Yorkshire in-laws were fond of saying. 'I suppose I could give you a short résumé.' Hell, I thought. I should have refused.

'Good' he quickly interjected, sensing my hesitation. 'That would be greatly appreciated.'

My wife came in with the coffee and admired the vase.

'How did you find the wine when you were in Prague?' asked Serc.

'Quite pleasant,' I replied noncommittally.

'I'll bring some round.'

The conversation became general and drifted round to my time at STC.

All those engineers who were made redundant must be scattered throughout the industry now,' commented Serc.

'Yes, but we meet up now and again. We have a reunion every year.'

'Do you know anyone in Marconi?'

'Possibly. Most of the engineers got very good positions in other companies.'

'We are probably going to buy some new communication equipment from them,' said Serc. 'I am unable to get the technical background we have asked for. You might be able to get it more easily.'

'I doubt it,' I responded.

Nevertheless he kept phoning me and asking when it would be convenient to see me again and deliver the wine. I pleaded pressure of work. He called and delivered the wine anyway. The information I had given him on radio altimeters was already typed. It was a copy of the paper I had presented at the Moscow symposium I had attended while at STC. All I had to do was alter the title and remove the introduction. I had given him nothing that was not already in the public domain. He did not realise this and was delighted with it. However I was uneasy. I needed advice. I could go to MI5 if I knew where it was. Was it really that serious? I knew our local MP was on various Defence and Aerospace committees, so I spoke to him.

'Leave it to me,' he said.

And so it was that I received a call from a Mr Johnson asking if I could meet him in his office in Whitehall, which turned out to be in the Foreign Office building. I filled in the obligatory pass and was escorted to an office on the third floor. Mr Johnson, if that was really his name, which I doubt, was exactly the sort of up and coming young man you would expect to see in the upper echelons of the Civil Service: well groomed, probably an Oxbridge graduate, very articulate with an easy manner. He introduced me to Carol, the other occupant of the office. In retrospect she looked remarkably like a younger version of Mrs Rimington, who later became Head of Intelligence – maybe she was.

'As you may have gathered,' said Mr Johnson, 'we are very interested in your contact. We would like you to keep in touch with us and let us know every time you meet him or have any contact with him. You can phone Carol on this number,' he added, handing me a slip of paper. 'If you ask for Carol you will be put straight through to her or one of her colleagues.'

'What about the information I am getting for him?' I asked.

'I gather it is pretty innocuous. He will probably escalate his demands. Keep in touch.'

'What about receiving gifts?'

'Gifts are OK, but do not accept money, not that I think he will offer it, not for the moment anyway. You were wise to get in touch with us. Recently there was an engineer employed in the defence industry who allowed himself to drift into a compromising position, discussing his work while being wined and dined. Fortunately we were able to get to him before any real damage was done and point out the error of his ways. He was genuinely upset by the way he had been led on. Although he had signed the Official Secrets Act we did not prosecute, but he lost his job, which was a pity because he was very good at it.'

Carol asked me if Serc had phoned me in the evening recently.

'Yes; three times when I was out,' I said.

'That explains why he kept nipping out to a phone box while at a reception,' she said. 'Let me know everything, even phone calls.'

Serc phoned again to ask me out to lunch. He also asked if I could find out who was developing infrared surveillance equipment for aircraft. He then said he would like me to meet a colleague who might have some useful information for me. He would be arriving at Heathrow in the morning, so perhaps I could meet him at the pick up point at terminal two and we could go to the terminal restaurant for lunch. I guessed that this meeting might be an offer of financial reward.

I rang Carol to tell her that I would like to withdraw from any further activity in connection with our Czech friend. The original contact had been exclusively to promote legitimate business for the company. It was now quite clear that none was forthcoming and I was really too busy to bother with his ridiculous requests. Also I did not like the sound of a new contact.

'I don't blame you,' responded Carol, adding that they would like one more meeting with me to tie up a few loose ends. She said that they would come over to me this time.

I would never have been a success as an undercover agent. I could not even make proper contact with my contacts. We had agreed to meet early for lunch. I arrived at the appointed time and as there were only two ladies in the dining room I waited in the reception area. They did not appear. I returned to the dining room to tell the headwaiter that I had booked for three but the other two hadn't turned up. 'Oh yes they have,' he said, 'They are waiting for you over there.' He gestured towards the two ladies as one of them turned towards me. It was Carol. Much embarrassment all round. Carol explained that Mr Johnson had moved to pastures new and introduced me to her colleague. We had a very enjoyable lunch.

I told them of my final meeting with Serc. I had tried to ring him back to cancel our meeting, but he was not available at the embassy. I had no alternative other than go to the airport and offer my apologies, pleading pressure of work, which was true. As I approached the terminal I saw Serc waiting.

'Look, I am sorry, I can't make it for lunch due to pressure of work and I won't be able to help you any further either. I haven't been able to obtain the recent items you asked for.' He went red with anger, glared at me and without saying a word slammed the door and stalked off back into the terminal building. Carol laughed.

'Surely,' I said, 'he must be the most crude and unsubtle of agents. Do you think he was a decoy to lead you away from some more devious activity?'

'No,' replied Carol. 'It is all pretty crude really; not at all like you read in books.'

CHAPTER XIII

GPWS - Ground Proximity Warning System

Thanks mainly to the BEA and MoD contracts we made steady progress for five or six years. We also looked for new products. One such came through the good offices of the Civil Aviation Authority.

A private aircraft owner and holder of a private pilot's licence who also happened to be a senior design engineer at Hawker Siddely had designed his own lightweight 360 channel VHF communication radio. As is the way in a fraternity of enthusiasts, other private pilots in the flying clubs got to hear about it and wanted similar equipment, commercial equipment being beyond their means. In consequence he found himself with a mini home industry, assembling sets with the help of his wife and family and selling them one at a time to his colleagues in the flying clubs. The scale of the operation was getting out of hand and was becoming an embarrassment to the CAA who had to approve each one individually for flight worthiness.

The CAA asked if we could take over production. As a CAA

approved organisation we could release them for sale as CAA approved items. There looked to be a market among private fliers for such equipment so we took it on.

It took nearly eighteen months of engineering development and drawing office work to make it suitable for small-scale batch production, whereby all the piece parts that made up the unit were fully defined and the build standard was up to CAA requirements. Maximum sales were about 50 a year, and at around £400 each this had little impact on our turnover. An updated 720-channel unit with an improved frequency synthesiser followed with a modest increase in sales.

Another activity was a distribution franchise for specialised switches approved for use in aircraft. There are several hundred switches in a large airliner; unseen switches which are activated when wheels are lowered, cargo and passenger doors opened etc. The mark up was small. New stock had to be paid for within 30 days, whereas some of the customers kept us waiting for 90 days. The larger companies such as GEC and Plessey were the worst. Their business was valuable and they knew it, but a delay in payment could mean us paying huge interest charges. Sometimes it was possible to turn the tables. A production manager would ring up full of venom because his order has been outstanding for several weeks.

'Don't you realise I have a big production order held up for just this one small item?'

'Until your company settles our outstanding account we will not be releasing any further items.'

'That's nothing to do with me. I have a responsibility to deliver on time and I need those switches. You get large orders from us. How dare you let us down?'

'If it's that important, surely it's worth your company settling our account. Then you can have your switches.'

'Can't you realise that in a company of this size I can't tell accounts what to do?'

'You will have to try.'

Eventually we would receive our cheque and send the switches by Securicor, which the nearly demented buyer would agree to pay for.

There must be a better way of doing business. I believe there is now legislation to prevent this sort of thing. What a pity it needed legislation.

With the completion of the BEA autoland and GPWS projects and the upgrading of the ministry equipment we were reliant on new projects such as a micro-switch franchise and light aircraft communication set, together with the traditional maintenance activity. It was obvious that this was not enough. The problem was compounded by political events outside our control.

It started with the Wilson government. With the need to spend more and more on his socialist programme he needed to cut expenditure in areas where it wasn't too obvious. There was a gradual fall off of traditional MoD repair work. Orders for spares went down to zero. This led to a ludicrous situation when the navy were engaged in the so-called cod wars with Iceland. A vessel was sent in for maintenance work. It had on board a helicopter. The helicopter had on board a piece of our equipment. The equipment was sealed against ingress of moisture by a close fitting cover, which needed replacing. There were no spare covers in naval stores. We were asked to make some in a hurry. Special machining operations had to be sub contracted, but sub contractors are reluctant to break off existing work for a small special order that has not been scheduled to take its turn, often with a lead-time of several weeks. Hence the need for advance ordering.

The ship was held in port until the cover was delivered direct to the dockside. 'For the want of a nail...'

When Margaret Thatcher took over things should have got better. I was discussing this with Bob one morning.

'You obviously haven't seen the paper this morning,' he said.

'No, I leave before it comes. What's new?'

'There has been a four hundred million overspend by the MoD, and Mrs T has called for a six-month moratorium on all new defence spending. There will be no more orders for six months.'

'That will finish us,' I exclaimed. In effect it did. It was the beginning of the beginning of the end. I was an ambitious engineer and I wanted to succeed. I wanted the firm to succeed and I wanted my family to reap the benefits. When we first started our ex-colleagues cast envious eyes at us as we were doing rather better by comparison. I was chasing success, but as it turned out the pot of gold was at the other end of the rainbow.

Good sound engineering practice was not enough. Marketing and research had to play a part, and we had sought a hole in the market to fill. We had tried behind the iron curtain without any real success. My sortie into our own government for new business showed promise initially. Funding for research and development was oriented towards the larger companies. I was encouraged therefore, when an assistant director mentioned that there was an interest in a low-cost lightweight radio altimeter and if I could demonstrate a viable prototype he would look favourably on funding further development. Since we were not being funded for the viability study I had to produce a lab model more or less part time, often in the evenings after the business of the day was done. This demonstrated our ability to produce what was required.

One fundamental point had been overlooked, although past experience should have given the necessary warning that in the ministry the left hand does not know what the right hand is doing. Furthermore it often does not wish to know. This is nothing new. Neville Shute in his autobiography *Slide Rule* made the same observation, going so far as to suggest that the *R100* airship disaster was the result of such an attitude.

When my original contact at the ministry moved on to another department his successor did not show the same interest in the development. He would not sanction the further engineering required for a full evaluation at A&AEE at Boscombe Down, the MoD trials establishment, even though viability had been confirmed by bench measurements at the Royal Signals and Radar Establishment at Malvern. It was easier and more comfortable to continue with the more expensive and heavier equipment already in service and manufactured under licence from an American company, to whom a licensing fee had to be paid in dollars.

For every successful invention there are probably several hundred failed ones, and they do not always fail due to flaws in the design. I had already developed a height sensor for use with the mandatory Ground Proximity Warning System for passenger carrying aircraft. Such a system was too complex and expensive for light aircraft. I adapted a simple radio altimeter adding forward-looking Doppler that measured the rate of approach of ground ahead. By resolving distance and rate of approach a hazardous situation could be established. The system showed promise but the company was under capitalised for carrying through such an ambitious development programme to its conclusion. There were ongoing contracts, in particular the Ministry post-design services contracts which resulted in much needed financial return. These had to take priority. I still believe that the light aircraft GPWS unit would have been viable and could save lives, but it was shelved.

There seemed no future. I left the sinking ship.

CHAPTER XIV

Consultancy

My first consultancy contract was with my old firm looking after some of the ongoing projects. I secured a job as chief engineer at Techtest Ltd. I was chief of two other engineers, two technicians and half a draughtsman. There followed one of the most satisfying periods of work, equal only to part of my time at STC. The location was at an old manor house. My domain was the loft area, the roof beams secured by wooden pegs. The view from my office window was rural with the Welsh hills in the background.

The core product of Techtest was test equipment for the in-situ checking of aircraft antennas, but also produced an NPL-designed primary standard of attenuation capable of resolving to a ten thousandth of a dB. It embraced several disciplines – the growth of a copper cylinder waveguide to internal tolerances of nanometres, the movement of a transmitting source on air bearings controlled by a zero backlash worm and wheel in steps resolved by a laser interferometer to a resolution of half a wavelength of red light. This was sold at a fabulous price to test houses worldwide.

My last project was the design of a UHF/VHF, AM/FM receiver for locating distress beacons. I had a credibility problem (nothing changes) until I took the marketing manager up in a helicopter. Operating at a test frequency the pilot flew directly over a beacon that had been placed in a hidden location.

I then found a new career in teaching. My first four years were with GNVQ (General National Vocational Qualification) courses. I had no time to prepare for my first appointment and was in for something of a culture shock.

I had already had some experience of teaching in a further education college after I qualified as a chartered engineer in 1958. The demise of my appointment at Techtest prompted me to try again. I was to teach three days a week at a college in Isleworth, foundation, intermediate, and advanced in applied science and mathematics for engineering.

I turned up on Monday morning not knowing what to expect. I picked up the register from the staff room and was directed to the appropriate classroom. I was greeted by a room of about twenty sullen looking students. They had been without a teacher for several months. Their previous teacher was on indefinite sick leave.

I decided to start by defining the basic SI units. It was difficult to get a response from them. They appeared to know nothing, which in fact turned out to be the case. This was true of all levels including to a large extent the advanced level. I had to start with basic numeracy. However as time went on we made some progress. I set them some homework, but had very little returned the following week. That which was returned was quite good.

I now realised the problem. There were one or two in each class who were eager to learn. I don't know to this day why the rest were there. Between the two extremes there were those who would like to get a qualification, but were unwilling to do anything out of class. At the bottom end they expected a qualification simply by turning up. The trouble was that the last category disrupted the lesson so badly that those that could benefit were penalised. I thought that my lack of teaching experience was to blame. This may have partly been the problem, although when discussing it with the permanent staff I found they acknowledged that it was a common complaint.

Why was this? My memories of teaching 45 years ago are that the pupils all responded, worked with a purpose and all attempted homework each week. Now the ethos is different. Has social engineering produced too many new further education colleges and are they too large? The funding is dependent on the number of students. This creates a conflict. There appeared to be a tendency not to be too stringent in the selection of pupils. Bring them in. Increase the funding.

Then there is retention, which is one of the masses of data that has to be sent to the relevant education authorities, and it must look good; so hold on to all the pupils, disruptive or not. Progression is another brownie point for the college. Progression of students to the next academic year together with pass rates is also important data by which the college is assessed and funded.

Second year classes will have pupils 'progressed' from the previous year, although unable to cope in that year. They take up the time that should be spent on the pupils that have the capability to succeed. So the pupils who could benefit are held back and don't benefit, the teachers are frustrated, and the tax-payers' money is wasted. The advice from my colleagues was that you just had to live with it and do the best you could.

After two years I moved college and took BTEC classes. This was better, but posed another problem. How on earth can you assess a student's mathematical ability by portfolio alone? Group work is fine during the learning process. The poorer students benefit by working together with more able students; but without a conventional exam, perhaps taken together with the student's portfolio, how can you separate the good from the not so good?

At the age of 80 I am now an agency lecturer teaching radio communication to final year HND students. Starting with my early association with the cat's-whisker crystal radio, this covers my 70-year relationship with radio communication and associated disciplines. I teach what I have done all my life. It is an appropriate conclusion.

APPENDIX

Everyman's Guide to the Evolution of Radio and Electronics

In the Beginning

It has been said that the 'invention' of radar changed the world. In that radar stemmed from radio technology it would, perhaps, be more correct to say that the invention of radio changed the world. Either way, radio and radar technology spawned electronics, which today affects almost every activity of our lives.

This has come about almost exclusively in the last century. It will be noticed that I put invention in inverted commas. Neither radio nor radar was invented. They evolved. My life from the early crystal sets to the electronics of today has paralleled this evolution.

It was in 1873 that Clerke-Maxwell, a Scottish mathematician, investigating the phenomena of magnetic induction (demonstrated earlier by Faraday), established that a changing electro-magnetic field sent out transverse electro-magnetic undulations having characteristics similar to light. Thus, like light, they could be reflected – the basis of radar!

Some fifteen years after Clerke-Maxwell's findings, Heinrich Hertz, a German physicist, demonstrated transmission of electromagnetic waves in the laboratory, using a spark gap connected to an electrical source through a coil of wire. Reception was sensed in a separate similar coil – the first transmission and reception of radio waves! Hertz also demonstrated that electro-magnetic waves travelled at the speed of light and could be reflected in a similar manner. This was not surprising since light itself is an extremely high frequency electro-magnetic wave. Hertz confined his experiments to the laboratory.

It was left to Marconi at the turn of the century to appreciate the possibilities of the transmission of electro-magnetic waves for worldwide communication. 'Imagination is as important as knowledge!' He started experimenting on his father's estate in Italy with the spark gap technique used earlier by Hertz. By progressively evolving this technique he was able to improve the detectable range. The main breakthrough was the use of a vertical wire connected to one end of the coil and a grounded metal plate at the other end – the first use of an aerial. By this method he was able to increase the detectable range to over 2km. He further demonstrated at this early stage that metal reflectors could concentrate the electro-magnetic energy into a beam. Detection at the receiver was by a coherer, loose iron filings in a glass tube that would 'cohere' under the influence of electro-magnetic waves.

These early crude experiments convinced Marconi of the future potential for a system of wireless communication. However, little interest was shown in Italy. He went to England.

In 1897 he transmitted signals across the Bristol Channel, a distance of 14km. His demonstrations attracted the interest of the chief engineer of the Post Office and he achieved recognition worldwide. This led to the setting up of the Wireless Telegraph and Signal Company, later to become the Marconi Wireless Telegraph Company.

The spark discharger generated a wide range of electromagnetic frequencies. Spark gap transmitters operating in the same area would interfere with one another. It was the academics that solved this shortcoming. Sir Oliver Lodge – when head of the physics department at University College, Liverpool in 1897 – was also experimenting with the propagation of electromagnetic waves. He used parallel plate capacitors in conjunction with the inductance coil associated with the spark generator to resonate at a particular frequency dependent on the values of the inductance and capacitor; hence tune the transmission and reception to a specific frequency. This enabled channel spacing between different transmitters. Marconi quickly made use of this tuning arrangement. It became standard practice in all further transmissions.

Although the academic community was investigating the transmission of electromagnetic waves, it was still Marconi who was evolving transmissions over long distances. He confounded the theorists by receiving signals transmitted beyond the horizon. In 1901 he spanned the Atlantic.

Later, at a frequency of 37Khz (a wavelength of 8000m) he received messages in Buenos Aires transmitted from Ireland, a total distance of over 9,000km. For this he used a long horizontal aerial.

The ability to receive signals over the horizon prompted further investigation into the propagation of electro-magnetic waves by physicists. Oliver Heaviside, a British physicist and engineer, concurrently with American engineer called Kennelly, predicted that long wavelength electro-magnetic waves were tunnelled between the ground and a layer of ionised air in the upper atmosphere. Another British physicist Sir Edward Appleton received the Nobel Prize for physics in 1947 for his seminal work on establishing the nature of the upper atmosphere.

In the meantime Marconi was concerned with the practicality of the effect and continued investigations at high frequencies in the region of 30–300MHz: wavelengths of 1–10m. He found that at these frequencies he could get poor or negligible reception at a distance of a few hundred kilometres, while good reception was obtained over much larger distances. Furthermore the distant signals could fade to nothing at different times.

This led to Marconi and other researchers to conclude that the signals were 'bouncing' off the ionisation layers in the upper atmosphere, and the nature of the ionisation changed with time. The ionisation was caused by radiation from the sun, changing night and day and according to the time of year. This skip distance effect was studied and the best times and frequencies for particular transmissions published in tabular form.

This high frequency band was used for maritime and aircraft communication. Amateur radio enthusiasts also used it, and with their knowledge of skip distances were able to communicate across continents and make a useful contribution to the knowledge of skip distance effects.

Until the evolution of the thermionic valve, or electron tube as it was called in the US, signals had to be sent in Morse code; the transmission was either on or off.

The Thermionic Valve Era

There were many efforts made to improve on the 'coherer', a detector of radio waves, as the radiated electro-magnetic waves became generally known. It was the physicist and engineer Sir Andrew Fleming (not to be confused with the discoverer of penicillin, Sir Alexander Fleming) who made the breakthrough. Fleming had studied under Clerke-Maxwell at University College, London. He had also worked with Thomas Edison, inventor of the light bulb. He investigated further the 'Edison Effect.' Edison had noted that a hot filament of tungsten wire emitted electrons. Electrons, being negative, could be attracted to a metallic plate at a positive voltage relative to the filament. Fleming took this effect further by placing the filament and plate in a glass vacuum tube. The extra energy in the heated filament excited the mobile electrons in the filament to such an extent that they would escape from the surface of the metal in a similar way to steam escaping from hot water. The electrons, however, did not have sufficient energy to escape completely, therefore forming an electron cloud round the filament in the absence of a voltage on the plate. As soon as the plate was made positive relative to the filament an electron current would flow from the filament to the plate. A current through a diode can only flow in one direction. By analogy to a mechanical or hydraulic valve it was termed a thermionic valve, or just 'valve' for short. The electron-emitting electrode is termed the cathode and the plate the anode.

At about the same time as Fleming was developing the diode, other forms of rectification were being investigated; in particular H.H.C. Dunwoody in the US developed the crystal detector. This was abandoned for many decades in favour of the more predictable diode, although it was used by hobbyists such as myself at a later date, and still later was the only practical means of detecting microwave transmissions.

It is worth recalling here that a convention initiated prior to the understanding of electron flow indicated current flowing from the higher potential (positive) to the lower – opposite to the electron flow. The convention is retained.

The development of the diode changed the whole concept of radio reception. An alternating current varying at any frequency such as those received from radio transmissions could be converted to a direct current. Furthermore if the transmitter frequency were to be amplitude modulated at voice frequencies this amplitude modulation could be recovered at the receiver by suitably integrating the unidirectional current from the diode detector. Wireless telephony replaces wireless telegraphy! It is a logical step to amplitude modulate the transmitted carrier frequency with music as well as voice. However public broadcasting awaited the evolution of the triode valve.

In 1906, just two years after Andrew Fleming developed the thermionic diode, Lee De Forest found that by adding a thin wire grid between the cathode and anode it was possible to control the current flow by applying a negative bias voltage relative to the cathode. Furthermore a small variation in the bias voltage makes a large change in the electron current flowing from the cathode to the anode – it amplifies a signal applied to the grid.

The development of the triode amplifier had important ramifications on the way both radio transmitters and receivers operated. In the transmitter the output signal could be fed back to the input of the triode via a resonant circuit tuned to the desired transmitter frequency. Since the triode has a gain of greater than unity, self sustained oscillations are maintained at the desired frequency. These oscillations can be further amplified to feed power to the aerial for transmission. An audio signal applied to one of the latter amplifiers will amplitude modulate the radio frequency carrier. In practice a more sophisticated method of modulating the transmitter carrier frequency is used employing a diode matrix to mix the audio and RF carrier frequencies.

In the receiver the triode amplifies the radio frequency selected by the tuned circuit before rectification by the diode. The audio frequency separated from the transmitter carrier frequency can now be amplified by several triode valves to produce an audio frequency signal having the power to drive a loudspeaker.

In the UK the first public broadcast service, The British Broadcasting Company, was set up by radio manufacturers in 1922. It broadcast on a long wavelength of 1500m – 200 kilocycles per second (200kHz). This was the forerunner of the BBC, set up in 1927. The BBC longwave transmitter set up at Droitwich continued at 200kHz (a wavelength of 1500m) delivering 100 kilowatts to a vertical aerial having an electrical length of 380m (quarter wavelength). This could be received halfway round the world. Other lower power 'regional' transmitters were set up on the medium waveband (300kHz – 3MHz).

The first wireless set owned by my family around 1927 was typical of the early TRF receivers. It was reckoned to be portable, but the heavy batteries required meant it needed a strong arm to carry it. The input coil doubled as a frame aerial, wound round the inside of the cabinet. A ganged variable capacitor tuned the frame aerial and plate circuit of the radio frequency amplifier to the required frequency. The second triode valve used the control grid as a diode to form a 'leaky grid detector'. With no negative bias applied to the grid the positive half cycles of the radio frequency carrier wave cause the grid to take current. This current drives the coupling condenser to a negative voltage proportional to the amplitude of the carrier and hence follows the audio amplitude impressed on it at the transmitter. A resistor connected to the junction of the grid and coupling capacitor allows the negative voltage to leak away at a rate that will follow the level of the audio frequency but not the much higher carrier frequency. The plate current follows the audio voltage developed at the grid. This audio signal is now amplified in two following valve amplifiers to a level that is sufficient to drive the moving-iron loudspeaker.

In these early sets the volume was adjusted by a 'reactance control' at the radio frequency amplifier. The early triode valves had very little gain. The reactance control fed a limited amount of the amplifier RF output back to the input, adding to the received signal. If too much feedback was applied the amplifier would burst into oscillation. As this was coupled to the aerial a whistle would be heard in all the neighbours' sets.

The second family set was homemade. It was still a TRF set, but I used more advanced valves with additional grids to screen the anode plate from the grid to prevent oscillation at higher gains, and a third grid close to the plate connected to the cathode to repel secondary emissions of electrons from the plate when bombarded by electrons emitted from the cathode. I also used a mains transformer to supply the valve heater current and rectified mains to provide the high voltage DC supply for the valve anodes. It was not until around 1936 that we acquired a 'superhet'.

The Superheterodyne Receiver

The formal definition of superheterodyne is 'a form of incoming radio signal converted to an intermediate frequency by mixing with a locally generated signal to facilitate amplification, and the rejection of unwanted signals.' In other words the received radio frequency signal is mixed with a local oscillator signal in the receiver that is tuned to resonate at a fixed frequency difference to the received signal. The mixer is usually a diode or diode matrix. The output of the mixer is the sum or difference of the two input frequencies. Received signals at any frequency are converted to a single intermediate frequency by selecting with tuned circuits at either the sum or difference frequency at the mixer output. It enables stable amplification over a defined frequency band, giving superior rejection of unwanted out of band frequencies prior to detection of the audio frequency.

The superheterodyne receiver was initially developed in the US by an army communications officer, Major Edwin Armstrong. It remains the basis of broadcast and radio communication reception.

Frequency Modulation

One of the disadvantages of amplitude modulating the radio frequency carrier is the presence of atmospheric and manmade electro-magnetic noise. Little can be done to reduce atmospheric noise. Manmade noise from electrical machinery such as car ignitions and washing machines can be suppressed at source. There are mandatory requirements for the suppression of these forms of noise. Nevertheless some noise comes through if the wanted signal is weak. This takes the form of a background hiss in radios. The noise is primarily an amplitude-modulated signal. A frequency modulated transmitter maintains a constant amplitude of carrier signal. The audio signal modulates (deviates) the transmitter frequency of the constant amplitude transmittor.

At the receiver the audio signal is extracted by a frequency discriminator instead of an amplitude detector. Limiting the carrier signal to constant amplitude prior to the frequency discriminator eliminated any amplitude variations of the carrier. This frequency-modulated arrangement virtually eliminates background noise until the received signal is too weak to be limited in amplitude prior to the frequency discriminator. The majority of communication systems operating at VHF and above are frequency modulated systems. At frequencies below about 100MHz frequency modulation is not practical due to the percentage of the carrier frequency that would have to deviate, typically 10kHz above and below the carrier frequency.

Digital radio will eventually replace frequency-modulated systems. Digital techniques are used almost exclusively in mobile phones. The advantage of digital transmission and reception is the discreet difference between 1s and 0s. There is nothing in between, hence virtually no noise. In general the digits are defined by a discrete frequency shift between the 1 and 0 digits.

Television

Having achieved sound broadcasting it was not surprising that the radio industry as a whole began to turn their attention to the transmission of moving pictures. John Logie Baird had demonstrated television images as early as 1926. Although early BBC broadcasts initially used the Baird system, picture resolution and integrity suffered due to the use of mechanical scanning. Picture synchronisation between transmitter and receiver was difficult. The receiver had to be continually adjusted to prevent the picture drifting across the screen. There was very little room for improvement. To be 'first with the worst' in no way diminishes Baird's achievement in developing the initial television system.

The research and development facilities of the radio industry did not seek to evolve or improve the Baird system, but to develop an all-electronic system to give greater flexibility and stability. The system developed by EMI and the Marconi Company was chosen as the standard by the BBC and has remained the basic system prior to the recent development of digital electronics.

Broadly speaking each picture frame is made up of 625 scanning lines. At the transmitter the scene is focused onto a photoelectric plate. A scanning electron beam detects the light level across the plate. This is the information that is transmitted at VHF frequencies. The modulation intensity is applied to the electron beam of the receiver cathode ray tube, which has the scan synchronised to the transmitter by synchronising pulses at the start of each picture frame. Flat screens transpose the picture scan onto a matrix of liquid crystals.

Radar

Clerke-Maxwell, when the transmission of electro-magnetic waves was only a mathematical concept, had appreciated that they could be reflected in the same way as light. Hertz had not just demonstrated electro-magnetic radiation; he had also indicated reflection characteristics. Marconi utilised the reflecting characteristics to enhance radiation in a particular direction, just as a headlight reflector focuses a beam. The headlight also reflects back the images of obstacles in the light beam.

The principle of radar was there for all to see, yet its application in the detection of aircraft came to Watson-Watt from an unrelated investigation into the possibility of using a powerful source of radiation as a death ray. At the same time it was noticed that aircraft flying in the vicinity of a television transmitter caused interference to reception.

Robert Watson-Watt (later Sir Robert) as superintendent of the Radio Department of the National Physics Laboratory, investigating the ionosphere, asked a junior scientific officer, Arnold Wilkins, to calculate the transmitted power needed to obtain measurable reflections from aircraft. The calculations showed that radio reflections from aircraft had sufficient energy to be detected by radio receivers.

France, Germany, Japan and the US had each in their different ways investigated the detection of aircraft from reflected electro-magnetic waves. Briet and Tuve in the US used a basically similar technique in 1926 to investigate the height of the ionised layers of the upper atmosphere.

It was only in Britain that the significance of the technique was realised at the highest level. Churchill, as chairman of the defence committee, appreciated its importance for early warning against an airborne attack. Money was allocated for further research under Watson-Watt. He chose a team of university academics and engineers from the radio industry to set up a research facility at Bawdsey Manor near Orfordness on the Suffolk coast. The team set about their work with enthusiasm and urgency.

The basis of the technique was the transmission of a short pulse of about four microseconds duration every two hundredths of a second. The pulse would trigger the start of a time base on a cathode ray tube. The time base would travel across the tube in 2,000 microseconds. The speed of electro-magnetic waves in free space is close to 186,000 miles per second. The distance from start of the time base to the received echo from an aircraft shows up as blip on the time base. The distance of the blip from the beginning of the time base indicates the time taken for the go and return signal, and hence range.

Bearing posed a problem. However if the range from two adjacent stations about ten miles apart were compared, the bearing could be found geometrically. This was cumbersome. A more practical method was to have two adjacent receiver aerials separated by a defined distance on the common transmitter tower, the phase difference measured by a goniometer giving an indication of bearing. More sophisticated methods could be used using common aerials for transmission and reception. Using a transmission frequency of 45MHz a chain of RDF (the original designation for radar) stations was established round the south and east coasts of Britain to provide early warning of the approach of enemy aircraft. The use of a 200MHz transmission frequency enabled a narrower beam width with aerials mounted on a rotating platform to give improved bearing information and coverage of low flying aircraft.

Having established the feasibility of ground-based radar the challenge was to develop airborne radar. Ground-based radar transmitters and receivers were built onto chassis mounted in nineteen-inch by six-foot racks. Dr Bowen, one of the original scientists who had developed the ground equipment, took up the challenge. Airborne equipment would need small light aerials. A 1.5m wavelength (a frequency of 200MHz) was a practical possibility. The early development models were mounted on 'breadboards' – literally five plyboards with the supply and ground rails top and bottom, the valve bases screwed down along the middle; the components connected as appropriate. Sufficient transmission power was obtained by operating the transmitter output stage well above its rating for the short period of the pulse. Reception was by a modified television receiver.

Not long after the successful trials of ground radar, members of the team, flying at 10,000ft, received echoes from ships off the Suffolk coast at a range of several miles. Air to Surface Vessel (ASV) radar was born. Later, in the same year (1937) Wood and Bowen flying in an Anson aircraft picked up the aircraft carrier *Courageous* and associated warships on the cathode ray tube screen. It also identified Swordfish aircraft launched from the carrier – the first air-to-air detection. The struggle to get satisfactory air-to-air operation for aircraft interception (AI) was hampered by the resolution of target bearing using a wavelength in the region of one metre. This required the development of microwaves.

Microwave Development

Work on microwaves was being carried out in several countries including the US, primarily in the Bell Labs. The aim was to get down to 10cm (3,000MHz – 3GHz). It was operational at a wavelength of 3cm by the end of the Second World War.

The basic technology was fairly well established. Due to the short wavelength of the signal, conventional wiring could not be used. Connections had to be made by transition lines, co-axial cable or, for minimum loss, waveguides. The oscillator tube used almost exclusively in all the research locations was the klystron. The transit time of the electrons was correlated to a pair of resonant cavities to create bunching of the electrons at the resonant frequency of the cavities. The output from the final cavity fed back to the first cavity maintains oscillations at the resonant frequency of the cavities.

By its nature, the focusing of the electrons into a very narrow beam, the peak output power of the klystron was limited to a few hundred watts – insufficient for really effective radar. Air to ship radars kept to 1.5m pending a breakthrough in microwave power sources. Low power microwave radars were developed for aircraft interception, enabling night fighter closure at short range. The search for improved microwave technology was universal, but nowhere more urgent than at TRE in 1940, to combat the submarine menace to shipping in the Atlantic, and for night fighters to combat the change from daytime bombing to night bombing.

TRE worked in close collaboration with both the universities and industry. The Birmingham University research facility under Professor Oliphant was asked to look into an improved microwave source.

It is sometimes possible to upgrade an existing device. Two of the Birmingham research team, Howard Boot and John Randal, looked at an earlier development that had showed no promise – the magnetron. In its original form the magnetron produced less power than the klystron. However the geometry was different. The limitation of the klystron was the small size of the cathode necessary to emit a narrow electron beam. The magnetron had a central cathode surrounded by a circular anode. A rotating electron stream was created by a magnetic field interacting with the electrons emitted from the cathode. By adding several resonant cavities within the anode structure, similar to those of the klystron, the large rotating electron field couples to the cavities in the anode. Furthermore the copper anode allows heat dissipation needed for larger power operation. Their handmade model produced a power an order better than the klystron. Further development by GEC boosted the output to 15 kilowatts peak pulse power. Thus the cavity magnetron was born, and has a wide variety of applications such as modern microwave ovens, point-to-point microwave communication links and radio astrology. It solved the transmitter power requirements. Reception remained a problem.

The First Microwave Detector and Silicon Chip

Detection of microwaves had been a problem from the beginning. Thermionic valves were impractical at these short wavelengths. In the early days of 'wireless' the cat's whisker silicon crystal had been used for signal detection. When the BBC first started broadcasting on long wave it is estimated the up to 100,000 crystal sets were in use. As valve radios became available the crystal sets were considered obsolete and the cat's whisker was largely forgotten.

With the development of the first prototype microwave radars at TRE the quest for suitable detectors renewed interest in the silicon crystal detector. Why it rectified the problem was not fully understood. Dr Herbert Skinner was tasked to develop its potential as a microwave detector. He would adjust the spring wire to obtain a ratio of forward to backward conduction of at least five to one, then encapsulate it in a glass tube. BTH took over production, improving the connections and sealing the position of the wire with wax.

So important was the microwave detector for the production of microwave radars that it was agreed to call on the US to take on the development as part of a technological cooperative agreement. It transpired that the Bell Labs had been carrying out independent investigations relating to their pre-war interest in microwave communications. More than a million silicon diodes had been produced in the US by the end of the war.

The Bell Labs have always been in the forefront of communication technology and associated research and development. It was a natural development to investigate the mechanism of the cat's whisker. The US was now the major and virtually the only player in the post-war electronics revolution and its associated dramatic fall out, whilst Britain languished in a political change that eschewed enterprise and stifled initiative.

Cat's Whisker to Transistor

The resurgence of the cat's whisker twenty years after it was believed to be yesterday's technology created renewed interest into just what went on in the critical area where the tungsten or steel wire touched the crystal. The Bell Labs with their multi-disciplined teams were well placed to investigate the properties of silicon and similar semi-conducting material such as germanium. It became apparent that the level and nature of the impurities in the silicon played a part in determining the direction and level of current. The position of the cat's whisker determined which particular impurity predominated at the point of contact.

To investigate the matter further, very pure germanium was refined and the effect of different impurities was noted.

The atomic structure of elements was well understood by physicists. By deliberately adding an impurity (about 0.1 of 1 per cent) to the otherwise pure germanium having one less orbiting electron, the interacting lattice is deficient of electrons and thus positively charged – p-type germanium. Conversely an element added with one more orbiting electron than germanium has a surplus in the lattice – n-type germanium. Early development was done with germanium, as it was easier than silicon to purify and then dope with other elements.

Diodes continued to use a tungsten wire contact to the germanium. Control of the impurity resulted in predictable performance. Just as the thermionic diode was converted to a triode amplifier, could the semiconductor diode be transformed to an amplifier by the incorporation of a third electrode? Yes it could.

Researchers at Bell Labs placed two finely pointed wires very close together on a p-type germanium base. The current flow between the two points could be modulated by a control current applied to the base. This was the point contact germanium transistor; it was used in the first transistor radios. As with many discoveries the initial development is not necessarily the best solution.

William Shockley, who later had a special microwave diode named after him, investigated a new approach. He placed a thin wafer of n-type germanium between two pieces of p-type germanium. A small current applied to the n-type region could control the current flow between the two p-type regions. This was the pnp junction transistor. Either germanium or silicon could be used. An npn configuration behaved in the same manner by applying currents of opposite polarity.

Silicon junction transistors were the preferred option, as they were less susceptible to temperature changes. Texas Instruments produced the first silicon junction transistors. It was their earliest type of transistor, the 2N001, that I used for the prototype transistorised 'autoland' radio altimeter in the mid 1950s.

Other types of transistor followed the junction transistor. The most important was the field effect transistor (FET), and its derivative the MOSFET, so named due do the insulation of the doped silicon areas by metal oxide – Metal Oxide Silicon Field Effect Transistor.

Typically a substrate of n doped silicon is masked with silicon oxide. Very small adjacent gaps in the mask allow diffusion of p doping in the two areas. A thin metal electrode is deposited onto the small gap between the two p-doped areas, insulated from the substrate by the metal oxide. This forms a channel between the two p-doped areas and is termed the 'gate'. A negative bias on the gate electrode will control a current flowing between connections to the p-doped areas, referred to as the SOURCE and DRAIN electrodes respectively. The sauce gate and drain electrodes of FETs are broadly analogous to the emitter base and collector of junction transistors. The main advantage of the FET is lower current consumption.

Using successive photographic masking processes, several thousand transistors can be formed on a single silicon dice, interconnected by diffusion of gold between electrodes as appropriate. Automation of the process once set up enables integrated circuits to be produced by the million. Transistorised integrated circuits form the basis of digital and computer technology.

Digital Technology

Initially the transistor was conceived as an analogue signal amplifier primarily for communication systems. Its function in digital circuits is fundamentally different. In its analogue function the power supply has no part to play in its function so long as it provides sufficient power to enable it to operate and amplify the signal; i.e. the signal is independent of the supply voltage. In a digital circuit it is either on or off – it performs a switching function between the supply voltage, typically 2–5 volts and nominal zero (typically 0.2 volts). Thus there are just two states. There is a voltage or there isn't.

About 160 years ago the British mathematician George Boole developed a system of logic on the basis that something is or isn't – logic functions 1 or 0. Simple logic based on this premise has been in everyday use for years. For instance, consider the basic logic function of a remotely operated door that will only open if a key is turned and a button is pushed.

If the key isn't turned and the button isn't pushed – 0 and 0 – the door isn't opened – output function 0. If only one of the inputs are activated – 1 and 0 – the door isn't opened – output function 0. If both the key is turned and the button is pushed, 1 and 1, the door is opened – output function 1.

In another situation an automatic door is made to open if either one or another (or both) of two push buttons are operated to give function 1 at the output.

The above are examples of AND and OR logic.

The AND logic can be inverted such that it requires two 0s to give a 1 output. This is a NAND function. Similarly if one or both inputs are 0 to give a 1 output it is termed a NOR function.

The speed of the transistor switch can enable an output function as a result of several million input functions being satisfied – the basis of computer programming, all carried out at machine level by 1s and 0s. Furthermore several million bits of data can be stored in an electronic memory all in the form of 1s and 0s, albeit in 8, 16 or 32 bit 'words.' Why in groups or multiples of eight?

Consider the application of numbers using only 1s and 0s; the 1 has to be shifted to 10 to be equivalent to decimal number 2, just as 9 goes to 10 in decimal notation. Thus the digital numbers shift one place every times two – as decimal notation shifts every times ten. Four becomes 100, eight 1000, sixteen 10000 and thirty-two becomes 100000; hence thirty-three is 100001, thirty-four 100010 etc. Sixty-three is 111111 so sixty-four becomes 1000000, i.e. an additional 0 is added every time the number is doubled from the previous shift to the left. This is binary notation. In calculators all numbers are changed to binary notation for processing by Boolean logic and converted back to decimal notation at the output. Similarly computers store data in binary notation.

Simple logic such as the door example above could be executed using mechanical switching such as relays, as originally used for switching in telephone exchanges. Arguably the first digital programmable computer was used during the Second World War at the code-breaking establishment at Bletchley Park, based on a thesis written by the Cambridge mathematician Alan Turing, one of the leading Bletchley Park code breakers. Using 1,500 valves for binary switching it occupied a whole room. A second version had 2,500 valves. Today's laptops operate at a thousand times the data rate of the original valve equipment. It might be said that during the early 1940s Britain led the world in computer technology as well as radar techniques. After the Second World War the lead crossed the Atlantic.

The Microprocessor

The microprocessor is the heart of the modern computer. It performs the essential functions of the central processor unit (CPU), performing logical operations, storage of data in conjunction with random access memories (RAM) and read only memories (ROM) all in the form of 1s and 0s. An internal clock, a crystal oscillator as in a digital watch, controls timing. The oscillator is one of the few analogue circuits in microprocessors. The output is converted into digital timing pulses to control operational sequences.

Microprocessors have many other applications such as the control of domestic washing machine programmes, engine management systems, compact disks and remote control vehicles on Mars, or any data management system. Photographic techniques enable up to a million transistors to be interconnected on a single silicon substrate one inch square for a typical microprocessor.

Masers and Lasers

With the computer it could be said that electronics has reached its apogee. Further developments move in the direction of quantum electrodynamics. The development of the transistor at Bell Labs was carried out by scientists who were familiar with quantum mechanics as well as microwave technology.

One such scientist was Nobel prizewinner Charles Townes. He established that by radiating a material from a microwave electro-magnetic source of the same or similar frequency as the molecular resonance of the material, a quantum change in the energy level of the atoms would occur, releasing energy in the process. Typically ammonia gas irradiated with a microwave source of its molecular resonance of 28.87GHz would emit energy at the same frequency. Furthermore the frequency of the stimulated radiation would be extremely stable. The name given to this effect was Microwave Amplification by Stimulated Emission of Radiation – MASER.

The beauty of the stimulated microwave emissions from a molecular source was the purity with regard to noise content and frequency stability. Signal/noise ratio is the limiting factor for receiver sensitivity. Use of a laser amplifier enabled sensitivities two orders better than previous low-level microwave amplifiers.

Light is a source of electromagnetic waves. It followed that a similar stimulation of light waves to that of microwaves was possible. It was Townes with a Bell Lab colleague who proposed Light Amplification by Stimulated Emission of Radiation – the LASER. As with the maser the stimulated emission from molecular resonance results in very pure coherent light of a single frequency. There is negligible divergence. Depending on the technique and material used very intense beams can be produced having pulse powers of several thousand kilowatts.

The most common and probably the most important laser is the semiconductor diode laser. By using appropriate doping at the *pn* junction (typically gallium arsenide) stimulation is achieved from a small current from an emf of 2 volts, enabling it to be used in conjunction with conventional transistor circuitry and incorporated into an integrated circuit. It can also be modulated by the current supply at a rate of up to 20GHz. This makes it an ideal source for data transmission by fibre optics.

Radio Astronomy

At the end of the Second World War many of the academic staff that had been recruited from the universities to TRE, the main engine of advances in electronics, in particular microwave technology, returned to academic life. Some applied the new-found techniques to explore further into the universe, beyond the range of optical observation. Early pioneers of this work were Sir Bernard Lovell, who set up the huge radio telescope at Jodrell Bank, Dr Bowen, the pioneer of air to air radar used his experience to set up radio astronomy research in Australia and Sir Martin Ryle, who as a result of his astronomical work at Cambridge University, became Astronomer Royal.

The outcome of this pioneering work is well known. A new window on the universe was opening. To extend this knowledge further, better signal/noise ratios were required at the receiver. The improved sensitivity and resolution of MASERS enabled the detection and, by the use of spectrometry, the nature and temperature of the gas clouds in outer space.

The Laser Interferometer

Unlike white light, which contains the whole colour spectrum, molecular stimulated emission is at one precise frequency. Typically red light has a wavelength of around two thirds of a micron.

If a reflected laser beam is phase compared to the original source it will reinforce the source when in phase. When 180° out of phase it will detract from the combined source. Thus for every half wavelength there will be a step change in the combined light intensity. This step change will occur for a movement of the reflector of each quarter of the wavelength of the laser source. Thus changes in distance can be resolved to less than a micron. Each step change in intensity can be detected by a photo-diode. The sum of steps gives change of distance. The rate of change of steps gives velocity.

The Atomic Clock

The multi-disciplined teams working at the Bell Labs on spectrometry and the development of the maser realised that the frequency stability of molecular resonance provided a highly stable measure of time.

At the eighteenth international conference on weights and measures in 1967 the international unit of time was redefined in terms of the caesium standard. The second was defined as 'the duration of 9,192,631,770 periods of radiation corresponding to the transition between hyperfine levels of the ground state of the caesium 133 atom.'

Historically the Royal Observatory at Greenwich had defined the standard of time based on the time taken for the earth to circumvent the sun. It followed that the primary standard should continue there. It is now situated at the new site at Herstmanceau in Sussex.

Fibre Optics, Microwave Links and Satellite Communication

Traditionally communication was by wire or radio link. As the volume of communication traffic has increased, wire link capability has shown its limitations – in particular, attenuation over long distances and a limitation of the data that can be transmitted by coaxial cable.

Light is reflected from a mirror surface with very little loss. Light striking pure glass at a shallow angle is totally reflected. Light beamed down a thin fibre tube of glass will travel along the tube by total reflection from the walls. Using a laser source and selected glass fibres, data rates of several hundred million bits per second can be transmitted for distances of up to ten miles without repeaters; it is ideal for internet interconnections.

At microwave frequencies the wavelength is short enough to beam transmission in a particular direction with parabolic reflectors in a similar manner to directing light beams. The high frequency of microwaves enables the transmission of a large band of frequencies containing many channels of high-speed data. As with light microwave, links are usually limited to line of site communication and a bit beyond due to refraction in the upper atmosphere. Microwave transmissions beamed to satellites fitted with reflectors extend the effective line of site enabling transcontinental and ultimately, by the use of several satellite links, global communication.

Global Positioning System – GPS

GPS satellites have orbits at heights of 12,000 miles. Their positions in orbit are determined by triangulation of the distance from fixed points on the ground or from other satellites whose positions are known. The distances are established by measuring the time delay of the transmitted signal from the known positions. Atomic clocks are used to measure the time delays and hence the distances.

A GPS receiver on the ground will receive relative delay times from transmissions of three or four satellites and from that information calculate its position relative to them and hence ground position. The resolution of the distances is a function of the electromagnetic wavelengths of the transmitted signals – typically a fraction of a metre for microwave transmissions. Summing all possible errors the theoretical accuracy of GPS is of the order of one metre.

Compact Disks

A common application of the laser is in conjunction with the compact disk. The digital data is applied to the metalised surface of the disk by a moderately powerful laser beam which pits the track when pulsed on at a defined data rate. The spiral track has a width of the order of eight micro-metres – about a fortieth of the width of a human hair.

The track is read by means of a semiconductor laser focused on the track, which differentiates the reflected beam from the burnt or unburnt track. Tracking data is also written onto the disk to maintain the laser head above the track. The data from the disk is encoded in a microprocessor.

Index